CLIMATE CHANGE ECONOMICS

Commemoration of Nobel Prize for William Nordhaus

World Scientific Series on Environmental, Energy and Climate Economics

Series Editor: Richard S J Tol (*University of Sussex, UK*)

Published

Vol. 2   *Climate Change Economics: Commemoration of Nobel Prize for*
         William Nordhaus
         edited by Robert Mendelsohn

Vol. 1   *Climate and Development*
         edited by Anil Markandya and Dirk T G Rübbelke

Volume 2

World Scientific Series on
Environmental, Energy and Climate Economics

CLIMATE CHANGE ECONOMICS

Commemoration of Nobel Prize for William Nordhaus

Editor

Robert Mendelsohn
Yale University, USA

World Scientific

NEW JERSEY · LONDON · SINGAPORE · BEIJING · SHANGHAI · HONG KONG · TAIPEI · CHENNAI · TOKYO

Published by

World Scientific Publishing Co. Pte. Ltd.

5 Toh Tuck Link, Singapore 596224

USA office: 27 Warren Street, Suite 401-402, Hackensack, NJ 07601

UK office: 57 Shelton Street, Covent Garden, London WC2H 9HE

British Library Cataloguing-in-Publication Data

A catalogue record for this book is available from the British Library.

World Scientific Series on Environmental, Energy and Climate Economics — Vol. 2
CLIMATE CHANGE ECONOMICS
Commemoration of Nobel Prize for William Nordhaus

ISBN 978-981-124-768-2 (hardcover)
ISBN 978-981-124-769-9 (ebook for institutions)
ISBN 978-981-124-770-5 (ebook for individuals)

For any available supplementary material, please visit
https://www.worldscientific.com/worldscibooks/10.1142/12569#t=suppl

Desk Editor: Sandhya Venkatesh

Typeset by Stallion Press
Email: enquiries@stallionpress.com

FOREWORD

I used to argue that the Nobel Prize for environmental economics should go to William Baumol, William Nordhaus, and Martin Weitzman; Baumol for laying the foundations of environmental policy design; Weitzman for his work on policy design, green accounting, nature conservation, and discounting; and Nordhaus for his work on integrated assessment modelling and climate change. It was not to be. Baumol passed away and Weitzman, the environmental economists' environmental economist, was passed over. Nordhaus, the macroeconomists' environmental economist, shared the 2018 Nobel Prize with Paul Romer, who considered dynamic externalities of a different kind in growth models.

Nordhaus' Nobel Prize is well-deserved. In 1975, Nordhaus was the first economist to study greenhouse gas emissions and for 15 years, he was practically the only one, melding the optimization techniques designed by Alan Manne, the valuation methods developed by Ralph d'Arge and the social cost-benefit analysis pioneered by John Krutilla into what we now know as integrated assessment. When I started working on climate change, almost two decades after Nordhaus, I still had to explain to people what is anthropogenic climate change and why economists and everyone else should care about it. That is hard to imagine now that everyone has an opinion on the matter.

Nordhaus was a true pioneer. He almost single-handedly created a whole new field of economics, one that studies the mother of all externalities, global in scale, ubiquitous in its causes, pervasive in its consequences, long-lived, uncertain, ambiguous, and inequitable between and within societies. In his Nobel lecture, Nordhaus rightly refers to climate change as "the ultimate challenge for economics". Climate economics is blossoming. Nordhaus published the first estimate of the social cost of carbon, the Pigou tax, in 1982. Since then, 163 (and counting) papers have re-estimated this. Nordhaus published the first version of the Dynamic Integrated Climate Economy model in 1992. It is still the workhorse model for many young and by now not-so-young-anymore economists interested in studying optimal climate policy.

It is therefore a great honour and pleasure to publish this book in the *World Scientific Series on Environmental, Energy and Climate Economics*. Robert Mendelsohn did an excellent job in putting together an edited volume of contributions by leading scholars in the field of climate economics, contributions that go beyond the obligatory

"Bill is a great guy" (he is) and pay true homage to an outstanding scholar, to whom we all owe an enormous intellectual debt.

Richard S.J. Tol
University of Sussex, UK
Series Editor,
World Scientific Series on Environmental, Energy and Climate Economics

PREFACE

This book commemorates William Nordhaus being awarded the 2018 Nobel Prize in Economic Science. Professor Nordhaus earned this distinction with his pathbreaking research integrating science and macroeconomics into a coherent tool to study climate change. Every economic-climate model built has been influenced by his efforts. Climate policy now and into the future will be guided by his insights.

Professor Nordhaus's early career led him to be uniquely positioned to tackle climate change. His initial research explored the macroeconomics of growth. What role did innovation play in growth? Did finite natural resources limit growth? His early work also included many models of energy. What role did individual fossil duels play supplying past energy and what role would they play into the distant future?

So, it is perhaps no surprise that Professor Nordhaus was one of the first economists to become interested in modeling climate change and its interaction with energy. His first efforts explored how an externality like carbon dioxide would disrupt the future use of fossil fuels. However, these models gradually became ever more complicated by taking on the critical details of greenhouse gases — the delayed temperature response; the long residence time of carbon dioxide; the growing marginal damage of rising temperatures; the heterogenous regional cost and benefit of climate mitigation; and the potential long term catastrophic consequence of warming.

Professor Nordhaus has made many contributions linking modeling and climate policy. He has written many articles on the use of a carbon price schedule (the social cost of carbon over time) as a tool to encourage both cost effective and efficient global mitigation. He has contributed to measuring the magnitude of the social cost of carbon and written about how quickly cost rises if only a fraction of the world's greenhouse gas polluters participated in mitigation. He has also evaluated political constraints such as bans on nuclear energy or carbon capture and storage; looked at how delaying starting mitigation has changed future mitigation options; and examined using trade restrictions to encourage reluctant countries to mitigate.

In addition to his extensive writing, Professor Nordhaus has also been an inspiration to other authors in the field as an example, as a mentor, and as a critic. His influence has encouraged others to tighten their arguments and engage in more ambitious efforts.

This book is a collection of the comments of eleven of his students, colleagues, and critics. The collection is an impressive set of contributions in their own right. But more than anything else, they demonstrate the breadth and depth of the effect William

Nordhaus has had on environmental economics in general and climate change in particular.

Robert Mendelsohn
Yale University

CONTENTS

© 2021 World Scientific Publishing Company
https://doi.org/10.1142/9789811247699_001

CHAPTER 1

ROLLING THE DICE IN THE CORRIDORS OF POWER: WILLIAM NORDHAUS'S IMPACTS ON CLIMATE CHANGE POLICY

JOSEPH E. ALDY[*,†,‡,§,¶] and ROBERT N. STAVINS[*,†,‡,‖]

*Harvard Kennedy School,
Harvard University, Cambridge, MA 02138, USA

†Resources for the Future,
Washington, DC 20036, USA

‡National Bureau of Economic Research,
Cambridge, MA 02138, USA

§Center for Strategic and International Studies,
Washington, DC 20036, USA
¶joseph_aldy@hks.harvard.edu
‖robert_stavins@harvard.edu

The seminal contributions of William Nordhaus to scholarship on the long-run macroeconomics of global climate change are clear. Much more challenging to identify are the impacts of Nordhaus and his research on public policy in this domain. We examine three conceptually distinct pathways for that influence: his personal participation in the policy world; his research's direct contribution to the formulation and evaluation of public policy; and his research's indirect role informing public policy. Many of the themes that emerge in this assessment of the contributions of one of the most important economists to have worked in the domain of climate change analysis apply more broadly to the roles played by other leading economists in this and other policy domains.

Keywords: Climate change policy; environmental economics; social cost of carbon; carbon tax; climate club; national income accounts.

1. Introduction

The core contributions of William Nordhaus to academic economics were summarized by the citation of his 2018 Nobel Prize "for integrating climate change into long-run macroeconomic analysis" (Nobel Foundation, 2018). But beyond those academic macroeconomic insights, Nordhaus has had significant influence on US and international public policies to address climate change. Identifying those contributions is more challenging, partly because such influences are more subtle. That is our purpose

---

¶Corresponding author.
This chapter was originally published in Climate Change Economics, Vol. 11, No. 4, December 2020, published by World Scientific Publishing, Singapore. Reprinted with permission.

in this paper — to identify how Prof. Nordhaus and his research have affected the path of climate change policy in the United States and around the world.

1.1. *The use of analysis in policy formulation*

Because academic influences on policy can be subtle, they are easily missed. For many years, economists and others have taken note of the influence — or the lack thereof — of economic analysis on real-world policymaking. The upshot of much of that work was summarized by Schelling when he surveyed potential influences across a range of policy areas — including abortion, crime, environment, healthcare, illegal drugs, national defense, and race relations — and observed "how little difference economic analysis appears to make in most important policies" (Schelling, 1997). However, economics and economists can be influential without leaving a trail of evidence. Sometimes, economists' greatest influence in government is by stopping bad ideas, as opposed to successfully promoting good ones (Schultze, 1996). But, within the environmental policy realm, a case can be made that the influence of economics, even if relatively modest, has at least increased — with fits and starts — over the past four decades (Hahn, 2000; Schmalensee and Stavins, 2019).

A useful metaphor about the use of analysis in the process of policy formulation, which we learnt from our colleague, Dutch Leonard, is of a "light bulb versus a rock".[1] Does the analysis provide illumination about a proposed policy, and thereby persuade policymakers to pursue said policy on its merits? Or does the analysis function as ammunition in a policy battle, used by those who already support a given policy (on its merits or for other reasons)? Evidence from the Council of Economic Advisers (CEA) suggests that the use of macroeconomic analysis frequently fits the "light bulb" function (Schultze, 1996; Stein, 1996), whereas microeconomic analysis is more likely to function as a "rock" (Schelling, 1997; Schultze, 1996).

Over time, Congressional testimony by academic economists has increasingly come to fit within the category of ammunition, whether for members of a Congressional committee who are already supporters or are already opponents of specific legislation (Devins, 2005). Likewise, administrations frequently contact academic economists when a new legislative proposal or administrative action is about to be rolled out, in order to develop a list of "objective academics" who can serve as validators when contacted by the press. Similarly, many academic economists have received press inquiries, wherein it becomes clear that the reporter wants a statement of validation for a particular point of view, not an objective assessment.

1.2. *Three pathways of impacts on climate change policy*

In reflecting on how economists have had impacts on the formulation of public policies, we identify three distinct pathways. First, there is *direct participation in the*

---

[1]For an examination of how an economic analysis, even if imperfect, can shine a light on climate change policy, see Aldy (2004).

policy world, such as via full-time government service or part-time service on a government committee. In the case of Bill Nordhaus, we review, in Sec. 2 of this paper, three engagements that fall into this first pathway: (1) Member, Council of Economic Advisers, 1977–1979; (2) Chairman, National Academy of Sciences (NAS) Panel on Integrated Environmental and Economic Accounting, 1996–1999; and (3) Chairman, NAS Committee on the Effects of Provisions in the Internal Revenue Code on Greenhouse Gas Emissions, 2011–2013.

The second of the three pathways is that of *directly influencing the formulation of public policies*, a pathway we believe to be considerably less frequent than commonly assumed. We find one significant example of this in the case of Nordhaus, and examine it in Sec. 3 of this paper: the explicit use of the Dynamic Integrated Climate–Economy (DICE) model by an interagency working group in the Obama Administration which developed quantitative estimates of the "social cost of carbon" (SCC) (IWGSCC, 2010, 2013, 2015, 2016).

The third and final pathway is by far the most common, and we characterize this pathway as *indirectly informing public policy*. This category turns out to be in some ways the most challenging to assess, because here the picture is clouded by the first half of the well-known proverb that "success has many fathers, but failure is an orphan". Subject to that caveat, in Sec. 4 of this paper, we examine six distinct ways in which William Nordhaus's economic research has indirectly informed climate change policy.

In the final section (Sec. 5) of this paper, we identify themes that emerge from this investigation — both generally and specific to climate change policy. This leads us to conclude with an appeal for modesty in thinking about the influence of economists on climate change policy.

2. Participation in the Policy World

An economist can share his or her views on policy without filter or interpretation by personally participating in policy debates. "Being in the room where it happened." "Holding the pen." "Clearing on the memo." These phrases represent ways that direct participation in the policy process can enable an economist's voice to be heard and influence an outcome. Our first example of Nordhaus's participation in the policy world — as a Member of the Council of Economic Advisers — is being part of the process, while the second example — chairing National Academy of Sciences panels — is providing advice to the process.

2.1. *Council of Economic Advisers*

On March 18, 1977, the United States Senate confirmed William Nordhaus as a Member of the CEA, which is comprised of a Chair and two Members, and advises the President on domestic and international economic policies. In contrast with other agencies and departments of the Federal government, which may have their own interests and agendas, the CEA has "the luxury of trying to discern what is in the best interest of the country and of providing that analysis and advice directly to the President" (Feldstein, 1992, p. 1224).

As a Member of the three-person council, Nordhaus's portfolio was necessarily broad, as noted in the 1979 Economic Report of the President: "Mr. Nordhaus has supervised international economic analysis and microeconomic analysis, including analysis of policies in such areas as energy, agriculture, social welfare, and oversight of regulatory reform activities" (United States President and Council of Economic Advisers, 1979, p. 173). This service predated meaningful US or international concern about climate change, but years later, Nordhaus's efforts on regulatory reform and oversight were of significance in the climate policy realm.

In 1977, while Nordhaus was a Member of CEA, President Carter created the Regulatory Analysis Review Group (RARG) and charged CEA with its leadership (Litan and Nordhaus, 1983). The RARG could review analyses of regulations and make recommendations to regulatory agencies. Agency rules — as opposed to analyses — were not subject to RARG review, however, and recommendations were not binding on regulators. Not surprisingly, CEA advocated for more rigorous analyses of the benefits and costs of regulations (Sabin, 2016).

The next year, President Carter issued Executive Order 12044, "Improving Government Regulations", which established the objective that regulations "shall achieve legislative goals effectively and efficiently".[2] The Executive Order required regulatory agencies to explain the need for their regulations; consider the direct and indirect effects; evaluate potential alternatives; solicit and respond to public comments; communicate rules in plain language; estimate reporting and record-keeping burdens; and develop a plan for retrospective evaluations. However, the Executive Order did not formally require benefit–cost analysis, and it failed to institutionalize a centralized authority responsible for implementing it. Nevertheless, in advocating for analyses of the benefits and costs of regulations, Carter's Executive Order and his CEA set the stage for subsequent administrations to require such assessments from their regulatory agencies, beginning with President Reagan's Executive Order 12291 and President Clinton's Executive Order 112866 (Litan and Nordhaus, 1983; Kerry Smith, 1984; Arrow *et al.*, 1996).[3]

In Carter's CEA, Nordhaus expressed concern about the potential inflationary impacts of regulations and advocated for a regulatory budget (Sabin, 2016). This has long been a controversial proposal to impose a social cost budget on regulations akin to a spending budget on government appropriations. Of course, limiting the aggregate costs of regulations, without consideration of their societal benefits, undermines the role of benefit–cost analysis in identifying policy options that increase social welfare.

---

[2]Refer to Section 1 of the Executive Order, https://www.foreffectivegov.org/sites/default/files/regs/library/eo12044.pdf.
[3]The emerging role of SCC in monetizing the benefits of reducing carbon dioxide (CO_2) emissions in recent regulations — addressed in Sec. 3, below — reflects this legacy of promoting the explicit consideration of the benefits and costs of proposed regulatory actions.

A memorandum from CEA Chair Schultze to President Carter in advance of the 1979 State of the Union address followed up on a meeting of the President, Vice-President, Schultze, and Nordhaus on establishing rational regulatory priorities (Schultze, 1979). The CEA Chair acknowledged disagreement among White House staff regarding the idea of a regulatory budget. In his State of the Union address, Carter avoided the issue altogether.[4]

All in all, Carter's CEA and Nordhaus's service as a Member thereof had profound long-term effects on public policy — in the climate domain and many others — by initiating thinking that led directly to the use of Regulatory Impact Analyses (RIAs) of all major Federal regulatory proposals. By the year 2000, 20 of 28 countries in the Organization for Economic Development and Cooperation had implemented requirements for such RIAs (OECD, 2002).

2.2. *National Academy of Sciences*

In 1979, Nordhaus departed CEA after two years of service, the norm for CEA members. But he continued to inform the policy process through another vehicle for academic experts: committees convened by the National Academies. Soon after leaving government, Nordhaus served on an early *ad hoc* committee of the National Academies focused on the economic and social aspects of climate change (National Research Council, 1980), more than a decade before the Rio de Janeiro conference that produced the United Nations Framework Convention on Climate Change (UNFCCC).

For the next two decades, Nordhaus focused mainly on his academic work, with direct and indirect climate policy implications, as we discuss in Secs. 3 and 4. But in 1998, Nordhaus returned to direct engagement with the policy world, co-chairing an NAS committee convened to study ways of accounting for natural resources and the environment in the National Income and Product Accounts (Nordhaus and Kokkelenberg, 1999). The National Academies published the committee's report as *Nature's Numbers*. The report examined the history of national accounting and augmented accounting, suggesting that it is important to retain conventional, core accounts, but to develop additional, supplemental accounts that capture the output and value of non-market natural services, even if defining the scope and measurement of those augmented accounts presents challenges.

In their assessment of the valuation literature, the committee noted the opportunities and methods for accounting for climate change benefits in valuing carbon sequestration through forestry activities — an issue relevant to a variety of land-use policies targeting climate change, such as multilateral programs that promote reduced deforestation and increased afforestation in developing countries and tree-planting efforts in

---

[4]Executive Order 12291 issued by President Reagan in 1981 included a reference to a regulatory budget, but the Reagan Administration did not pursue it. In 2017, President Trump issued Executive Order 13771, "Reducing Regulation and Controlling Regulatory Costs", which called for regulatory cost budgets, and effectively slowed the development and promulgation of new regulations, including those that may address climate change.

domestic policy. They also noted how Clean Air Act benefit–cost analyses had failed to monetize the benefits of reducing carbon dioxide emissions, an issue addressed subsequently through the application of the social cost of carbon (which we address in the next section). The report included guidelines for accounting for nonrenewable and renewable natural resources and environmental pollution.

In 2013, the National Academies published the report of another Nordhaus-chaired committee, *Effects of US Tax Policy on Greenhouse Gas Emissions*, convened at the request of the Department of the Treasury (Nordhaus *et al.*, 2013). This study identified the tax provisions with the greatest effect on emissions that contribute to global climate change. It focused on energy-related tax expenditures and excise taxes, biofuels provisions, and broad-based tax expenditures. The committee recognized the limitations of existing models and scholarship, and made recommendations for future research to fill these gaps. In addition, the report highlighted the inefficiencies in the existing tax code in terms of its impacts on carbon dioxide emissions, both through fossil fuel provisions that may increase emissions as well as the high costs per ton of emissions reduced through some renewable energy and energy efficiency tax provisions.

In both cases, the Nordhaus-led committees made recommendations to improve public policy to better account for the adverse impacts of pollution. Indeed, each reflected well-understood insights from economics about how appropriately pricing pollution could alter behavior and investment and thereby increase social welfare. For a variety of political reasons, the committees' recommendations to account for the natural environment in Federal statistics, as well to reform the tax code, were not adopted.[5]

3. Direct Influence on Public Policy

In addition to service in government positions at CEA and on NAS committees, Nordhaus carried out some of his most influential work while back at Yale. Economists' voices can carry through to public policy when their research directly informs key elements of its design, evaluation, or implementation. This has been true in one important case with Nordhaus's research, the use of his path-breaking Integrated Assessment Model (IAM) of climate change by government analysts for the purpose of developing empirical estimates of the marginal damages of a unit of CO_2 emissions, the so-called SCC.

3.1. *The DICE model and the social cost of carbon*

IAMs combine scientific relationships about emission flows, atmospheric pollutant stocks, temperature impacts, with key economic relationships to summarize the effects

---

[5]Muller *et al.* (2011) illustrate an application of environmental accounting within a national income accounting framework, consistent with the *Nature's Numbers* recommendations. Their application focuses on local air pollution and carbon dioxide damages, but they acknowledge that no government statistical agency had explicitly linked such damages to specific industries.

of climate change on human welfare. The first to incorporate economics in a systematic way was Nordhaus's DICE model, which has played a critical role in both scholarly and policy endeavors by characterizing the implications of a changing climate.

Consideration of the SCC, the estimated economic damage due to the emission of one ton of CO_2 into the atmosphere in a specific year, originated in Nordhaus's scholarship in 1982 (National Academy of Sciences, 2017). While this work — and many subsequent papers — came well before meaningful policy debates about mitigating CO_2 and other greenhouse gases (GHGs), these early models provided a foundation for quantitative assessment when those debates turned to real policy options.

The social cost of carbon can be used as an element in the design of climate change policy, such as in setting the tax rate for a carbon tax or informing the development of a cap-and-trade program. Also, government analysts can employ the social cost of carbon in their benefit–cost analyses of regulatory proposals that are expected to reduce CO_2 emissions (Nordhaus, 2013). This latter application of the SCC has been the norm in US energy and climate policy for more than a decade.

3.2. *The social cost of carbon arrives in government*

In 2008, a Federal court required the Department of Transportation to revise a fuel economy regulation because it was "arbitrary and capricious" by failing to monetize the benefits of reducing CO_2 emissions.[6] This created demand for SCC estimates. It is common for regulatory agencies to include nonmonetized benefit categories in their RIAs (Aldy *et al.*, 2020), but the existence of numerical estimates (using DICE model calculations) precluded efforts to avoid monetizing CO_2 emission reduction benefits for major energy and climate regulations. Some of the initial efforts by the Department of Transportation to include SCCs relied on syntheses of the literature by Tol (2005, 2008), which were influenced by Nordhaus's SCC estimates from several years.

In 2009, the Obama Administration formed an Interagency Working Group on the Social Cost of Carbon (IWGSCC) to review the scientific literature on climate change damages and provide guidance for monetizing such damages in RIAs (IWGSCC, 2010). Federal agencies began to use a preliminary set of SCC estimates based on an interim review of the literature, with explicit discussion of Nordhaus's DICE model as a key input to this work, along with two other IAMs — the FUND model (Tol, 1997) and the PAGE model (Hope *et al.*, 1993).[7]

---

[6]Center for Biological Diversity v. National Highway Traffic Safety Administration, United State Court of Appeals, Ninth Circuit, 538 F.3d 1172 (9th Cir. 2008).

[7]See, for example, the Department of Energy final rulemaking "Energy Conservation Program: Energy Conservation Standards for Refrigerated Bottled or Canned Beverage Vending Machines" (74 Federal Register 44914, August 31, 2009), which references both Nordhaus's work and the interim guidance from the interagency working group.

The IWGSCC (2010) issued guidance to agencies on the social cost of carbon, based on modeling runs conducted by government staff with the three IAMs. In this initial report and in subsequent technical updates (2013, 2015, and 2016), the working group generated hundreds of thousands of SCC estimates, using a range of discount rates, estimates of future economic growth, population growth estimates, and climate sensitivity assumptions. The primary SCC for Regulatory Impact Analyses reflected the mean SCC across these models and modeling assumptions (with a 3% discount rate).

In 2010, the IWGSCC's primary SCC estimate for 2015 was $28 per ton of CO_2.[8] As the modelers updated their models, the government updated its SCC estimates. In 2013, the working group adjusted the SCC to $45 per ton CO_2, with a modest revision down to $43 per ton in 2015 and 2016. Nordhaus's own updated estimates of the SCC increased in a similar fashion.[9]

The SCC has been used to monetize damages across three presidential administrations, but has it had a significant impact? Hahn and Ritz (2015) reviewed more than 50 regulations over 2008–2013 with monetized benefits associated with reducing CO_2 emissions, based on the SCC. Although monetized emission reduction benefits comprised a significant fraction of total monetized benefits for some rules, overall there was little evidence that inclusion of such monetized benefits changed the sign of net social benefits, and, in only a few cases did including the SCC affect the choice among regulatory alternatives. On the other hand, Pizer *et al.* (2014) demonstrated that the choice of SCC can influence whether the monetized benefits of reducing CO_2 emissions under the (then-proposed) Clean Power Plan would exceed the estimated costs of regulation. If assessments of future CO_2 regulations are based solely on the targeted pollutant benefits, i.e., not including any co-benefits of reducing local air pollution, then the rule would deliver positive net social benefits only if the carbon reduction benefits estimated with the SCC exceed the monetized costs (Aldy *et al.*, 2020).[10]

Although the Trump Administration has turned away from considering a global SCC, various state governments have drawn on the work of IWGSCC to inform the design and evaluation of various state-level energy and climate change policies (Paul *et al.*, 2017). In addition, the SCC has played a meaningful role in the International Monetary Fund's (IMF) recent efforts to highlight and discourage fossil fuel subsidies around the world (Coady *et al.*, 2015, 2019). In its 2019 analysis, the IMF employed an SCC estimate of $40/t$CO_2$, citing Nordhaus (2017) and IWGSCC (2016), to monetize climate change externalities from fossil fuel use. Thus, the SCC — used

---

[8]All SCC estimates in this section are for a unit of CO_2 emissions in 2015, converted to 2018 dollars.

[9]Nordhaus (2008) reported an SCC estimate of $14 per ton of CO_2 for the year 2015, with Nordhaus (2014) increasing it to $23, followed by Nordhaus's (2017, 2019a) estimate of about $35 per ton of CO_2.

[10]In 2017, President Trump disbanded the Interagency Working Group with Executive Order 13868, "Promoting Energy Independence and Economic Growth" (82 Federal Register 16093, March 31, 2017). Since then, relevant regulatory agencies have continued to employ SCC estimates, but based on 3% and 7% discount rates, and focused exclusively on domestic benefits of reducing carbon dioxide emissions, resulting in empirical estimates in the range of $1–3 per ton of CO_2 emissions (Schmalensee and Stavins, 2019).

under various Federal regulatory authorities, state planning and regulatory actions, and IMF policy guidance and evaluation — emerged as a focal point in a decentralized, patchwork approach to climate policy.

4. Indirectly Informing Public Policy

The third pathway is indirect and often subtle, but is, in general, the most common way academic research influences public policy. We examine six ways in which Nordhaus's economic research has indirectly informed climate change policy[11]: (1) investigations of the balance between economic growth and climate protection; (2) applications and extensions of the DICE model, both by Nordhaus and others; (3) identification of the dynamically efficient time path for global GHG emissions; (4) promotion of cost-effective GHG emissions mitigation, particularly through carbon taxes; (5) recommendation of alternative approaches to international cooperation on climate change, in particular the use of international clubs of countries; and (6) participation in policy-relevant intellectual debates.

4.1. *Balance of economic growth and climate protection*

Two scientific realities of global climate change lead to important economic implications. First, greenhouse gases mix in the atmosphere, and so the location of emissions has no effect on impacts, rendering climate change a global commons problem. Hence, any jurisdiction taking action incurs the costs of its actions, but the climate benefits are distributed globally. Therefore, for virtually any jurisdiction, the climate benefits it reaps from its actions will be less than the costs it incurs (despite the fact that the global benefits may be greater — possibly much greater — than the global costs).

Second, greenhouse gases accumulate in the atmosphere (for more than 100 years in the case of carbon dioxide), and the climate impacts are a function of the stock (atmospheric concentration), not the flow (emissions). As a result, the most severe consequences of climate change will be in the long term. However, climate change policies and their attendant costs of mitigation will be upfront. This combination of upfront costs and long-delayed benefits presents great political challenges.

Together, the global commons nature of the problem plus its long time horizon make climate change a difficult political challenge, and explain why there are concerns in many countries that ambitious short-term climate policy actions could undermine economic growth. Indeed, Nordhaus was the first to suggest that instead of investing in early abatement strategies, it would be more efficient to invest in conventional capital, and then use the additional resources in the future to invest heavily in climate capital (Nordhaus, 1977), presumably both for mitigation and adaptation. The same analytical

---

[11]Our focus here is on how specific elements of Nordhaus's research have indirectly informed climate change policy. Some of these were likely linked with Nordhaus's decades of teaching Yale undergraduate and graduate students, as well as his authoring several books for general educated audiences (Nordhaus, 1994, 2008, 2013).

framework, but with updated estimates of relevant parameters, has gradually led Nordhaus to endorse much more ambitious time paths of emissions reductions (Nordhaus, 2019a).

4.2. *Applications and extensions of the DICE model*

The DICE model, first described by Nordhaus in a *Science* article in 1992, appears to have been the first IAM to combine basic climate science, biophysical impacts, and economics so that the benefits and costs of alternative time paths of climate change could be compared.[12] In the subsequent Regional Integrated Climate–Economy (RICE) model, Nordhaus and co-authors incorporated geographically specific information on emissions, damages, production, and consumption to compare alternative approaches to climate change policy, including market-based, cooperative, and non-cooperative approaches (Nordhaus and Yang, 1996).

Importantly, because Nordhaus made the DICE/RICE models publicly available (in GAMS code), other researchers were able to use and/or extend the models for their own analyses. As described in detail in Sec. 3 of this paper, the DICE model — together with the FUND and PAGE models — have been used to estimate the social cost of carbon (IWGSCC, 2010, 2013, 2015, 2016).

4.3. *The dynamically efficient time path of global GHG emissions*

Economists have long used the Kaldor–Hicks criterion (Hicks, 1939; Kaldor, 1939) to identify the most efficient policy to achieve some public purpose, by estimating the expected net present value (NPV) of alternative policies, with different time paths of actions and consequences. From early on, Nordhaus applied this thinking to identify preferred time paths of emissions (and emissions abatement), thereby distinguishing between an economically optimal path and one centered on possibly *ad hoc* quantity targets specified in terms of emissions, GHG concentrations, or temperatures (Nordhaus, 1977).

Nordhaus then used this standard to maximize the discounted net present value of utility, subject to a resource constraint, adding equations to represent activities that generate emissions into the atmosphere. He found that it was only in later time periods that it would be optimal to modify the energy system. In the short term, the optimal carbon price is low, but increases to a very high level by the end of the 21st century. He concluded, at the time, that the efficient program "requires little change in the energy allocation for 20 to 40 years" (Nordhaus, 1977). More recently, Nordhaus's analyses have led to much more ambitious time paths of recommended emissions reductions (Nordhaus, 2015, 2019a).

---

[12]At about the same time, Cline (1992) published his work with his independent global macroeconomic model, from which he recommended much more aggressive policy action. The early works of both Nordhaus and Cline demonstrate that "ideas matter" in focusing attention by scholars and the public, and that diffusion from ideas to policy actions can take decades or longer (Hahn and Stavins, 1992).

4.4. *Cost-effective GHG emissions mitigation*

For 100 years, since Pigou (1920), economists have endorsed the use of pollution taxes to address environmental problems. Earlier than others, Nordhaus advocated for pricing instruments to address climate change in the form of carbon taxes (Nordhaus, 1977). In order to correct for climate externalities of GHG emissions, he emphasized the need for control strategies which were both feasible scientifically and possible to decentralize, leading to the notion of a tax on the carbon content of fossil fuels. The importance of carbon taxes remained central to Nordhaus's work on climate change for over four decades (Nordhaus, 2007b, 2013, 2019a).

Countries around the world — including nearly all of the industrialized countries and large emerging economies — have launched or are in the process of launching national policies aimed at reducing their emissions of GHGs. Of the 169 Parties to the Paris climate agreement that have submitted specific pledges (known as Nationally Determined Contributions or NDCs), more than half refer to the use of carbon pricing in their NDCs. To date, some 51 carbon-pricing policies have been implemented or are scheduled for implementation worldwide, including 26 carbon taxes and 25 emissions trading systems (Stavins, 2020). Together, these carbon-pricing initiatives will cover about 20% of global GHG emissions (World Bank Group, 2019).

4.5. *Alternative approaches to international cooperation*

In the nearly three decades since UNFCCC was signed in 1992 in Rio de Janeiro, Brazil (United Nations, 1992), Nordhaus has consistently been critical of the approaches taken to international cooperation under the United Nations, including both the top–down approach of the Kyoto Protocol (United Nations, 1998; Nordhaus and Boyer, 1999) and the more recent bottom–up or hybrid approach of the Paris Agreement (United Nations, 2015; Nordhaus, 2020). Following the works by Barrett (2003), Victor (2006), and others, Nordhaus used the DICE model to examine an alternative, largely voluntary international regime — a climate club (Nordhaus, 2015).

Nordhaus has become the most prominent analyst and spokesman on behalf of the club approach, in which the club's members (countries) agree to harmonize emissions reductions using a carbon price, keep the revenue raised for themselves, and penalize nonparticipants by imposing a uniform tariff on all imports from them. Carbon tariffs would be incentive-compatible, rewarding the punisher, and incentivizing participation. Although potentially promising as a next step if the United Nations negotiations fail to produce meaningful progress, the analysis indicates that a tariff sufficient to stabilize the climate would lead to the collapse of the club (Barrett, 2018). This approach has not received serious attention in international climate negotiations, but it remains among the most prominent alternative approaches to the United Nations scheme. Whether the climate club idea, in general, and Nordhaus's analysis and advocacy, in particular, will eventually be influential and affect the direction of international policy is anyone's guess.

4.6. *Policy-relevant intellectual debates*

Two major debates among economists in which William Nordhaus had prominent roles stand out in terms of their policy relevance. One was about the role catastrophic risk plays in assessments of climate change policies. The other concerns the low discount rates used by Nicholas Stern in his analysis of optimal policy.

Beginning 10 years ago, Weitzman carried out theoretical analysis of how positive biophysical feedback loops could lead to uncertainty about the future damages of climate change that would be best characterized by a probability distribution of damages with fat tails, such as a Pareto distribution, rather than a conventional Gaussian (normal) distribution (Weitzman, 2009, 2011). The result is greater weight being given to catastrophic (but relatively small probability) outcomes, such as in net present value analyses of alternative climate change policies. In the extreme, Weitzman's "Dismal Theorem" indicates that under specific structures of uncertainty and preferences, the expected value of losses would be infinite, and so standard economic (NPV) analysis simply would not apply.

Nordhaus (2011) and Pindyck (2011) pushed back against these findings, in part by indicating that the conditions of uncertainty and risk-aversion under which Dismal Theorem is derived are very limited, as well as by noting the absence of any policy interventions in Weitzman's analysis. Nordhaus suggested that policy interventions, such as geoengineering, that can truncate a fat tail merit serious consideration. In a recent paper, Nordhaus applies his thinking to account for a catastrophic event by incorporating the melting of the Greenland ice sheet in the calculation of the social cost of carbon (Nordhaus, 2019b). He finds that the risk of disintegration of the ice sheet adds less than 5% to the estimated social cost of carbon, because the damages would occur much later than many other damages already incorporated in the estimated social cost.

The other climate-related intellectual debate in which Nordhaus has been a major participant involves the discount rate employed in the Stern Review (Stern, 2007; Nordhaus, 2007a), the prominent analysis of climate change and related public policy carried out for the government of the United Kingdom. Nordhaus's critique of that report centers on its use of a near-zero social rate of time preference, which Nordhaus maintains would mean that the savings rate would have been much greater than it has been historically. Capital markets, he argues, reflect how societies have chosen to distribute consumption across time, and on that and other bases, claims that Stern has underestimated the appropriate social discount rate. He notes that if transfers were allowed, using Stern's policy prescriptions would actually make the world worse off, yielding in net present value terms, negative net benefits. He also points out that with such a high value of future consumption, we would need to be willing to trade off a large fraction of today's consumption for only a small fraction of potential consumption in the distant future.

5. Conclusions

The seminal contributions of William Nordhaus to fundamental scholarship on the long-run macroeconomics of global climate change are abundantly clear. More challenging to identify are the impacts of Nordhaus and his research on public policy in this domain. In the short term, at least, more easily digestible symbols — for example, targets of 2°C maximum temperature change by 2100 or net zero emissions in 2050 — sometimes matter more for public policy than even the best, most rigorous scholarly research. This is hardly unique to this case, however. Many of the themes that emerge in our analysis are common to the roles played by other leading economists in this policy domain, and perhaps in some other policy domains as well.

The limited influence of economic analysis on real-world policymaking has long been noted, although within the realm of environmental policy, such influence has increased over recent decades. Whether as a light bulb or a rock, the work of economists such as Bill Nordhaus can affect policymaking via three distinct pathways: personal participation in the policy world; directly influencing the formulation of public policies; and indirectly informing public policy.

In general, personal involvement in the policy world — in Nordhaus's case decades ago at the Council of Economic Advisers and the National Academy of Sciences — typically provides the most transparent evidentiary trail. Although even here, it is clouded by the passage of time. Some of Nordhaus's most important contributions to climate change policy from this period — such as his support at President Carter's CEA of the use of benefit–cost analysis to assess proposed government regulations — came to fruition only years later, beginning in the Reagan Administration, with the first formal requirements for Regulatory Impact Analysis.

A second pathway, which many economists seem to think is important and even quite common, is that of directly influencing the formulation of public policies. When and if this occurs, the evidence ought to be abundant and compelling. So it is with the use of Nordhaus's DICE model as one of three IAMs employed by the Obama Administration to estimate empirically the social cost of carbon. That the SCC has itself been used repeatedly for benefit calculations in RIAs demonstrates this direct connection of Nordhaus's research and the formulation of public policies. However, it should not be surprising that we find only one example of this second pathway, because this pathway — we believe — is a rare one in the work of any environmental economist.

We find that the third pathway — indirectly informing public policy — is the most important one through which Nordhaus's work has had impacts on public policy. Reflecting on the work of other scholars suggests that this is broadly the case for academic economic research influencing public policy, at least in the environmental realm. But here the evidence can be difficult to discern, partly because "success has many fathers".

Reflecting on these three pathways reminds us of Schumpeter's trichotomy of the process of technological change (Schumpeter, 1942). Schumpeter distinguished three steps in the process by which a new, superior technology permeates the marketplace. *Invention* constitutes the first development of a new product or process. In the policy context, think of a new economic theory or method, such as Nordhaus's DICE model. Most inventions never develop into an *innovation* (accomplished only when the new product/process is commercialized, i.e., moves from the laboratory to the showroom floor). And most academic research, no matter how path-breaking it may be as pure scholarship, never makes its way into influence in the real world of public policy. Last in Schumpeter's three stages of technological change is diffusion, the process of gradual adoption of a product or process. Here the analogy to the policy world is clearest, where some academic economists may work full or part time within government, directly pressing for the adoption of some specific policy, whether the policy idea is attributed to them or others.

William Nordhaus is perhaps the most important economist to have ever worked on climate change and climate change policy. If we are correct that his influence in the policy world has been as subtle and often as indirect as we have indicated, then perhaps we all should be left with a sense of greatly enhanced modesty regarding our own contributions to public policy. Our assessment in this paper of the policy impacts of one of the most important economists to have ever worked in the environmental domain prompts us to issue an appeal to our colleagues in the profession for greater moderation overall when making claims about academic research influencing policy developments, whether specifically for climate change or more broadly in the realm of environmental, energy, and natural resource policy.

Acknowledgments

The authors acknowledge the excellent research assistance of Kristin McCormack and valuable comments on a previous version of the manuscript by Jason Bordoff, Robert Hahn, Arik Levinson, and Richard Schmalensee, but the authors are responsible for all remaining errors. Joe Aldy acknowledges the research support of Resources for the Future, BP, and the Belfer Center for Science and International Affairs.

References

Aldy, JE (2004). Saving the planet cost-effectively: The role of economic analysis in climate change mitigation policy. in *Painting the White House Green: Rationalizing Environmental Policy Inside the Executive Office of the President*, R Lutter and JF Shogren (eds.), pp. 89–118. Washington, DC: Resources for the Future Press.

Aldy, JE, M Kotchen, M Evans, M Fowlie, A Levinson and K Palmer (2020). Co-benefits and regulatory impact analysis: Theory and evidence from Federal air quality regulations. Prepared for the NBER Environmental and Energy Policy and the Economy Conference, May 21.

Arrow, KJ, ML Cropper, GC Eads, RW Hahn, LB Lave, RG Noll, PR Portney, M Russell, R Schmalensee, V Kerry Smith and RN Stavins (1996). Is there a role for benefit–cost analysis in environmental, health, and safety regulation? *Science*, 272(5259), 221–222.

Barrett, S (2003). *Environment and Statecraft: The Strategy of Environmental Treaty-Making*. London: Oxford University Press.

Barrett, S (2018). Choices in the climate commons. *Science*, 362(6420), 1217.

Coady, D, I Parry, N-P Le and B Shang (2019). Global fossil fuel subsidies remain large: An update based on country-level estimates. IMF Working Paper No. WP/19/89, Fiscal Affairs Dept., International Monetary Fund.

Coady, D, I Parry, L Sears and B Shang (2015). How large are global energy subsidies? IMF Working Paper No. WP/15/105, Fiscal Affairs Dept., International Monetary Fund.

Devins, N (2005). The academic expert before Congress: Observations and lessons from Bill Van Alstyne's testimony. *Duke Law Journal*, 54(6), 1525–1554.

Feldstein, M (1992). The council of economic advisers and economic advising in the United States. *The Economic Journal*, 102(414), 1223–1234.

Hahn, RW (2000). The impact of economics on environmental policy. *Journal of Environmental Economics and Management*, 39(3), 375–399.

Hahn, RW and RA Ritz (2015). Does the social cost of carbon matter? Evidence from US policy. *Journal of Legal Studies*, 44(1), 229–242.

Hahn, RW and RN Stavins (1992). Economic incentives for environmental protection: Integrating theory and practice. *American Economic Review*, 82(2), 464–468.

Hicks, J (1939). The foundations of welfare economics. *The Economic Journal*, 49(196), 696–712.

Hope, C, J Anderson and P Wenman (1993). Policy analysis of the greenhouse effect: An application of the PAGE model. *Energy Policy*, 21(3), 327–338.

Interagency Working Group on Social Cost of Carbon, United States Government (IWGSCC) (2010). Technical Support Document: Social cost of carbon for regulatory impact analysis under Executive Order 12866. February. Available at https://19january2017snapshot.epa.gov/sites/production/files/2016-12/documents/scc_tsd_2010.pdf. Accessed on 9 July 2020.

IWGSCC (2013). Technical Support Document: Technical update of the social cost of carbon for regulatory impact analysis under Executive Order 12866. November. Available at https://obamawhitehouse.archives.gov/sites/default/files/omb/assets/inforeg/technical-update-social-cost-of-carbon-for-regulator-impact-analysis.pdf. Accessed on 9 July 2020.

IWGSCC (2015). Technical Support Document: Technical update of the social cost of carbon for regulatory impact analysis under Executive Order 12866. July. Available at https://obamawhitehouse.archives.gov/sites/default/files/omb/inforeg/scc-tsd-final-july-2015.pdf. Accessed on 9 July 2020.

IWGSCC (2016). Technical Support Document: Technical update of the social cost of carbon for regulatory impact analysis under Executive Order 12866. August. Available at https://www.epa.gov/sites/production/files/2016-12/documents/sc_co2_tsd_august_2016.pdf. Accessed on 9 July 2020.

Kaldor, N (1939). Welfare propositions in economics and interpersonal comparisons of utility. *The Economic Journal*, 49(195), 549–552.

Kerry Smith, V (ed.) (1984). *Environmental Policy under Reagan's Executive Order: The Role of Benefit-Cost Analysis*. Chapel Hill, NC and London, UK: The University of North Carolina Press.

Litan, RE and WD Nordhaus (1983). *Reforming Federal Regulation*. New Haven, CT: Yale University Press.

Muller, NZ, R Mendelsohn and W Nordhaus (2011). Environmental accounting for pollution in the United States economy. *American Economic Review*, 101(5), 1649–1675.

National Academy of Sciences (2017). *Valuing Climate Damages: Updating Estimation of the Social Cost of Carbon Dioxide*. Washington, DC: The National Academies Press.

National Research Council (1980). Letter from the *ad hoc* study panel on economic and social aspects of carbon dioxide increase. April 18. Climate Research Board, National Academy of Sciences, Washington, DC.

Nordhaus, WD (1977). Economic growth and climate: The carbon dioxide problem. *American Economic Review*, 67(1), 341–346.

Nordhaus, WD (1982). How fast should we graze the global commons? *American Economic Review*, 72(2), 242–246.

Nordhaus, WD (1992). An optimal transition path for controlling greenhouse gases. *Science*, 258(5086), 1315–1319.

Nordhaus, WD (1994). *Managing the Global Commons: The Economics of Climate Change*. Cambridge, MA: The MIT Press.

Nordhaus, WD (2007a). A review of the Stern review on the economics of climate change. *Journal of Economic Literature*, 45(3), 686–702.

Nordhaus, WD (2007b). To tax or not to tax: Alternative approaches to slowing global warming. *Review of Environmental Economics and Policy*, 1(1), 26–44.

Nordhaus, WD (2008). *A Question of Balance: Weighing the Options on Global Warming Policies*. New Haven, CT: Yale University Press.

Nordhaus, WD (2011). The economics of tail events with an application to climate change. *Review of Environmental Economics and Policy*, 5(2), 240–257.

Nordhaus, WD (2013). *The Climate Casino: Risk, Uncertainty, and Economics for a Warming World*. New Haven, CT: Yale University Press.

Nordhaus, WD (2014). Estimates of the social cost of carbon: Concepts and results from the DICE-2013 R model and alternative approaches. *Journal of the Association of Environmental and Resource Economists*, 1(1–2), 273–312.

Nordhaus, WD (2015). Climate clubs: Overcoming free-riding in international climate policy. *American Economic Review*, 105(4), 1339–1370.

Nordhaus, WD (2017). Revisiting the social cost of carbon. *Proceedings of the US National Academy of Sciences*, 114(7), 1518–1523.

Nordhaus, WD (2019a). Climate change: The ultimate challenge for economics. *American Economic Review*, 109(6), 1991–2014.

Nordhaus, WD (2019b). Economics of the disintegration of the Greenland ice sheet. *Proceedings of the National Academy of Sciences*, 116(25), 12261–12269.

Nordhaus, WD (2020). The climate club: How to fix a failing global effort. *Foreign Affairs*, May/June. Available at https://www.foreignaffairs.com/articles/united-states/2020-04-10/climate-club. Accessed on 9 July 2020.

Nordhaus, WD and JG Boyer (1999). Requiem for Kyoto: An economic analysis of the Kyoto Protocol. *The Energy Journal*, 20, 93–130.

Nordhaus, WD and EC Kokkelenberg (eds.) (1999). *Nature's Numbers: Expanding the National Economic Accounts to Include the Environment*. Panel on Integrated Environmental and Economic Accounting, Committee on National Statistics, Commission on Behavioral and Social Sciences and Education, National Research Council. Washington, DC: National Academy Press.

Nordhaus, WD, SA Merrill and PT Beaton (eds.) (2013). *Effects of U.S. Tax Policy on Greenhouse Gas Emissions*. Committee on the Effects of Provisions in the Internal

Revenue Code on Greenhouse Gas Emissions; Board on Science, Technology, and Economic Policy, Policy and Global Affairs; National Research Council of the National Academies. Washington, DC: The National Academies Press.

Nordhaus, WD and Z Yang (1996). A regional dynamic general-equilibrium model of alternative climate-change strategies. *American Economic Review*, 86(4), 741–765.

Nobel Foundation (2018). The Sveriges Riksbank Prize in Economic Sciences in Memory of Alfred Nobel 2018. Available at https://www.nobelprize.org/prizes/economic-sciences/. Accessed on 9 July 2020.

Organization for Economic Development and Cooperation (OECD) (2002). *Regulatory Policies in OECD Countries: From Interventionism to Regulatory Governance*. OECD Reviews of Regulatory Reform. Paris, France: OECD Publishing.

Paul, I, P Howard and JA Schwartz (2017). The social cost of greenhouse gases and state policy: A frequently asked questions guide. Report, Institute for Policy Integrity, New York University School of Law.

Pigou, AC (1920). *The Economics of Welfare*. London: Macmillan.

Pindyck, RS (2011). Fat tails, thin tails, and climate change policy. *Review of Environmental Economics and Policy*, 5(2), 258–274.

Pizer, W, M Adler, J Aldy, D Anthoff, M Cropper, K Gillingham, M Greenstone, B Murray, R Newell, R Richels, A Rowell, S Waldhoff and J Wiener (2014). Using and improving the social cost of carbon. *Science*, 346(6214), 1181–1182.

Sabin, P (2016). 'Everything has a price': Jimmy Carter and the struggle for balance in Federal regulatory policy. *Journal of Policy History*, 28(1), 1–47.

Schelling, T (1997). Why does economics only help with easy problems? In *Economic Science and Practice: The Roles of Academic Economists and Policy Makers*, PAG van Bergeijk, AL Bovernberg, EEC van Damme and J van Sinderen (eds.), pp. 134–148. Cheltenham, UK: Edward Elgar.

Schmalensee, R and RN Stavins (2019). Policy evolution under the Clean Air Act. *Journal of Economic Perspectives*, 33(4), 27–50.

Schultze, CL (1979). State of the Union: Language on setting regulatory priorities. Memorandum for the President of the United States, January 20, Jimmy Carter Library.

Schultze, CL (1996). The CEA: An inside voice for mainstream economics. *Journal of Economic Perspectives*, 10(3), 23–39.

Schumpeter, J (1942). *Capitalism, Socialism and Democracy*. New York: Harper & Brothers.

Stavins, RN (2020). The future of U.S. carbon-pricing policy. In *Environmental and Energy Policy and the Economy*, Vol. 1, pp. 8–64. Chicago, IL: University of Chicago Press.

Stein, H (1996). A successful accident: Recollections and speculations about the CEA. *Journal of Economic Perspectives*, 10(3), 3–21.

Stern, N (2007). *The Economics of Climate Change: The Stern Review*. Cambridge, UK: Cambridge University Press.

Tol, RS (1997). On the optimal control of carbon dioxide emissions: An application of FUND. *Environmental Modeling & Assessment*, 2(3), 151–163.

Tol, RS (2005). The marginal damage costs of carbon dioxide emissions: An assessment of the uncertainties. *Energy Policy*, 33(16), 2064–2074.

Tol, RS (2008). The social cost of carbon: Trends, outliers and catastrophes. *Economics: The Open-Access, Open-Assessment E-Journal*, 2, 1–22.

United Nations (1992). United Nations Framework Convention on Climate Change. Available at https://unfccc.int/process-and-meetings/the-convention/what-is-the-united-nations-framework-convention-on-climate-change. Accessed on 9 July 2020.

United Nations (1998). Kyoto Protocol to the United Nations Framework Convention on Climate Change. Available at https://unfccc.int/resource/docs/convkp/kpeng.pdf. Accessed on 9 July 2020.

United Nations (2015). Paris Agreement. Adopted December 12; entered into force November 4, 2016. Available at https://treaties.un.org/doc/Publication/CN/2016/CN.735.2016-Eng.pdf. Accessed on 9 July 2020.

United States President and Council of Economic Advisors (1979). *Economic Report of the President transmitted to the Congress*. January. Washington, DC: U.S. Government Printing Office.

Victor, DG (2006). Toward effective international cooperation on climate change: Numbers, interests and institutions. *Global Environmental Politics*, 6, 90–103.

Weitzman, ML (2009). On modeling and interpreting the economics of catastrophic climate change. *The Review of Economics and Statistics*, XCI(1), 1–19.

Weitzman, ML (2011). Fat-tailed uncertainty in the economics of catastrophic climate change. *Review of Environmental Economics and Policy*, 5(2), 275–292.

World Bank Group (2019). *State and Trends of Carbon Pricing 2019*. Washington, DC: The World Bank, Publishing and Knowledge Division.

© 2021 World Scientific Publishing Company
https://doi.org/10.1142/9789811247699_002

CHAPTER 2

WHAT THE FUTURE MIGHT HOLD: DISTRIBUTIONS OF REGIONAL SECTORAL DAMAGES FOR THE UNITED STATES — ESTIMATES AND MAPS IN AN EXHIBITION

GARY YOHE[*,‡], JACQUELINE WILLWERTH[†,§],
JAMES E. NEUMANN[†,¶] and ZOE KERRICH[†,‖]

*Economics and Environmental Studies, Emeritus
Wesleyan University 238 Church St. Middletown
CT 06459, USA

†Industrial Economics, Inc.
2067 Massachusetts Ave. Cambridge
MA 02140, USA
‡gyohe@wesleyan.edu
§jwillwerth@indecon.com
¶jneumann@indecon.com
‖zkerrich@indecon.com

The text and associated Supplemental Materials contribute internally consistent and therefore entirely comparable regional, temporal, and sectoral risk profiles to a growing literature on regional economic vulnerability to climate change. A large collection of maps populated with graphs of *Monte-Carlo* simulation results support a communication device in this regard — a convenient visual that we hope will make comparative results tractable and credible and resource allocation decisions more transparent. Since responding to climate change is a risk-management problem, it is important to note that these results address both sides of the risk calculation. They characterize likelihood distributions along four alternative emissions futures (thereby reflecting the mitigation side context); and they characterize consequences along these transient trajectories (which can thereby inform planning for the iterative adaptation side). Looking across the abundance of sectors that are potentially vulnerable to some of the manifestations of climate change, the maps therefore hold the potential of providing comparative information about the magnitude, timing, and regional location of relative risks. This is exactly the information that planners who work to protect property and public welfare by allocating scarce resources across competing venues need to have at their disposal — information about relative vulnerabilities across time and space and contingent on future emissions and future mitigation. It is also the type of information that integrated assessment researchers need to calibrate and update their modeling efforts — scholars who are exemplified by Professor

---

‡Corresponding author.

This chapter was originally published in Climate Change Economics, Vol. 11, No. 4, December 2020, published by World Scientific Publishing, Singapore. Reprinted with permission.

Nordhaus who created and exercised the Dynamic Integrated Climate-Economy and Regional Integrated Climate-Economy models.

Keywords: Climate change; damages; uncertainty; distributions; sector; region; United States; temperature; mitigation.

1. Introduction

In its Fifth Assessment Report, the Intergovernmental Panel on Climate Change (IPCC, 2014) identified changes to the climate system and socioeconomic development processes as key drivers of vulnerability, exposure, and hazard. It was understood that these processes would produce, or at least influence, material risk to human and natural systems over time and space. The Fourth Assessment Report (IPCC, 2007) had previously reported in its *Synthesis Report* that "responding to climate change involves an *iterative risk management* process that includes both *mitigation and adaptation* and takes into account *climate change damages*, co-benefits, sustainability, equity, and attitudes toward risk (emphasis added)". It was then immediately clear that investment in either mitigation or adaptation depended upon efficiently processing information about the observed and projected consequences of climate change as well as their relative likelihoods. This conclusion is still *the* rigorous foundation of robust response analysis — decisions based on changes in social and natural characteristics that have been detected and quantified from historical data and then, per-haps, attributed to climate change and its anthropogenic sources.

IPCC (2007) also produced a rigorous approach to attributing observed changes in global and continental mean temperatures to anthropogenic sources (for example see Figure SPM.4 in IPCC 2007). Authors reported results from an ensemble of climate models that were run with and without anthropogenic forcing. Bifurcations of 5th to 95th percentile ranges were tracked and reported to support confidence in attribution at various points in time. That is to say, the 5th percentile of trajectories that included anthropogenic forcing were higher than the 95th percentile of trajectories that ignored human influences. This achieved a very high confidence bar for asserting that the two post-bifurcation distributions of warming could not have been produced from the same (no anthropogenic forcing) underlying potential reality. Here, we apply this same approach (bifurcations of the 5th to 95th percentile ranges) to assess significant dif-ferences in outcomes between the climate scenarios explored in this work.

We are keenly aware that information valuable to mitigation and adaptation deci-sions with time horizons shorter than nearly a century is scarce. This is especially true for information about risks that are tracked in time between now and the year 2100. Fawcett *et al.* (2015) is typical of front-line publications that focus readers' attention on temperature distributions at the end of this century. There, an enormous number of emissions scenarios of heat absorbing gases were calibrated to produce projected distributions of equilibrium global mean surface temperature (GMST) for the year 2100. Papers like that, and there are many, offer increasing information about the end of the century, but they offer little if any information for years between 2020 and 2100.

In addition, few of them distinguish the difference between transient and equilibrium temperature scenarios; and fewer calibrate transient temperatures along alternative mitigation pathways. Perhaps the most concise and comprehensive review of economic cost estimates and approaches is Auffhammer (2018).

NRC (2010) became an exception to this pattern when it reported that observing X degrees of transient warming one year after another in the 21st century means that the planet has committed itself to long-term warming that will likely converge to $2X$ degrees. Nonetheless, seven years later, papers like Hsiang *et al.* (2017) were still estimating damage functions against equilibrium temperature gradients but frequently without reference to time.

Many contributions to the Special Report on Global Warming of 1.5° Degrees by the Intergovernmental Panel on Climate Change (IPCC, 2018a) worked to fill this gap. They compared the relative values of trajectories that limited GMST to 1.5°C warming instead of 2.0°C. Since most of the action in that comparison will happen between 2030 and 2070, these contributions depended on either (1) interpolations of transient temperature trajectories for the two limiting GMST maxima and/or on (2) damage functions calibrated against static GMST values without regard to location or time. Recognizing time, of course, is a much more difficult problem. It means that regional damage functions that would be relevant to regional and national adaptation planners would have to recognize regional variations in surface temperatures over time.

This paper was written to fill, at least partially, some of these significant gaps in our ability to project potential damages (consequences) from climate change and their likelihoods by sector incrementally over space and time along an assortment of possible emissions futures — exactly the sort of information that is necessary to design and implement effective forward-looking investments in climate-focused adaptation. We produce and display internally consistent and entirely comparable distributions of increasing regional temperature trajectories over time. Temperature changes were calibrated against the average global mean temperature over the 20-year climate era from 1986 to 2005. We then produced and compared regional damages from climate change across 15 sectors in the contiguous 48 states for which damage functions driven by regional temperature, or regional sea level rise in the case of coastal property damages, have been estimated. They are characterized and portrayed along four different emissions pathways for each of the seven regions that span the lower 48 states.

Section 2 presents brief descriptions of methods and procedures before we turn to reporting results in Sec. 3 (regional mappings of distributions of damage estimates for four emissions scenarios and qualitative interpretations by sector). Section 4 is devoted to estimates of total sectoral damages for the 48 contiguous states, and Sec. 5 explores implications for comparing the value of achieving 1.5° and 2.0°C mitigation targets — economic results that persistently display little or no statistical difference across 15 sectors and 7 regions. We close with some discussion of content and context with reference to Professor Nordhaus in Sec. 6.

2. Methods and Procedures

2.1. *Spatial and temporal dimension of the analysis*

Temperature projections and damage estimates were projected for four 20-year climate eras from 2020 to 2100. It is widely thought that it takes at least two or three decades to reliably distinguish changes in climate from inter-annual variability.[1] We chose these 20-year eras between 2020 and 2100 because climate change, at least as reflected by changes in GMST, is speeding up. In our analysis, these four eras are identified by their midpoints: 2030 (2020–2039), 2050 (2040–2059), 2070 (2060–2079) and 2090 (2080–2099). The analysis was conducted for seven regions covering the contiguous US. These regions were identified in the Third (US) National Climate Assessment (NCA3, 2014) and modified only slightly in the Fourth iteration (NCA4, 2018) — we adopted the more recent regional definitions.

2.2. *Characterizing distributions of transient GMST for four emissions scenarios*

Emissions scenarios from Yohe (2017) served as the foundation for our point of departure. The four emission scenarios were identified by their conventional names in the UNFCCC negotiations: that is, by "1.5 Degrees", "2.0 Degrees", "3.0 Degrees", and "BAU" scenarios, respectively. The "Business as Usual" scenario was drawn from Fawcett *et al.* (2015); it is consistent with IPCC emissions pathways labeled RCP8.5 (IPCC, 2008) and SRES A2 (IPCC, 2000); that is, it corresponds to a GMST change of about 4.8°C in 2100 relative to the preindustrial baseline. In relation to the 1995 era baseline used in the damage calculations, these four temperature scenarios ("1.5", "2.0", "3.0", and "BAU") correspond to changes from a 1995 era baseline of 0.85°, 1.35°, 2.35°, and 4.18°C, respectively.

These pathways were calibrated in decadal intervals through 2100 for the three temperature policy targets as well as a BAU baseline. They were ultimately derived from a fundamental conclusion from NRC (2010): a quasi-linear relationship exists between contemporaneous cumulative emissions of carbon dioxide and transient temperature change (a median estimate of 1.75°C of warming compared to a preindustrial baseline for every 1000 Gt of *carbon* contributed to cumulative emissions over any period of time). Uncertainty, here, is driven by uncertainty about the behavior of sinks in higher temperatures and by uncertainty about the sensitivity of the climate to external forcing. According to this relationship, the 95th percentile temperature for any emissions total is 70% above the temperature associated with median, while the 5th percentile temperature is 40% below the median in 2100.

The three median scenarios that constrain temperature increases below specified maxima relative to preindustrial levels were imported from Yohe (2017). They are "ideal" in the sense that they reduce emissions over time so as to maximize the

---

[1]The standard depends on how fast change is happening to collect and aggregate weather patterns into a coherent portrait of contemporaneous climate; see www.NASA.gov.

discounted logarithmic utility. That is to say, these median transient GMST pathways solve three parallel Hotelling-style exhaustible resource problems where limits on cumulative emissions derived from NRC (2010) serve as operating constraints on total emissions: 1720, 2570, and 4290 GtCO$_2$ through 2100 for maxima *relative to prein-dustrial temperatures* of 1.5°C, 2.0°C, and 3.0°C, respectively. The significance of the resulting scarcity rents for carbon emissions was expressed in terms of utility discount rates required to guarantee that the chosen emissions pathways in fact satisfy exactly their respective maximum temperature constraints along the *median scenario*. For reference, it is important to notice that the three temperature maxima are met exactly along only the median trajectories.

2.3. *Regional temperature trajectories for all seven regions driven by four discrete GMST pathways*

In the second step, all of the median GMST trajectories depicted were translated into regional temperatures pathways for seven NCA4 (2018) regions. The regional temperatures were estimated using the relationship between era regional temperatures and GMST found in the suite of GCM models underlying the original sectoral damage models: CanESM2 (von Salzen *et al.*, 2013), CCSM4 (Gent *et al.*, 2011), GISS-E2-R (Schmidt *et al.*, 2006), HadGEM2-ES (Collins *et al.*, 2011), and MIROC5 (Watanabe *et al.*, 2010). This group of five met criteria established in by the US Environmental Protection Agency, including a consideration of independence and quality for US projections, and ensuring a broad range of temperature and precipitation outcomes, as a group, over the continental US. The full set of selection criteria and results are described more fully in USEPA (2017) and its associated Technical Appendix.

The median GMST trajectories are also translated to regional sea level rise (the climate input used in the coastal properties damage estimation). First, we estimated GMST corresponding to the six sea level rise projections (Kopp *et al.*, 2014) used in USEPA (2017) by developing a relationship between global temperature and global mean sea level (GMSL) rise. We used global temperature changes from 1970 to 2100 for six GCMs under RCP8.5, and a GMSL rise projection for RCP8.5 derived by applying probabilistic weights (USEPA, 2017) to the six SLR scenarios and fit a quadratic, least-squares regression.[2,3] We inverted this function and applied it to the GMST trajectories to estimate a corresponding GMSL trajectory and to estimate re-gional sea level rise from the global measures. The global-to-regional adjustment relies on a relationship defined from the six sea level rise projections from Kopp *et al.* (2014) and the regional sea level rises for the same scenarios from Sweet *et al.* (2017).

---

[2]In addition to the five GCMs used in USEPA (2017) this relationship also considers GFDL-CM3 (Donner *et al.*, 2011), an additional GCM used in Sarofim *et al.* (2020) to enhance coverage of high-end temperature scenarios.
[3]The resulting function is $y = -0.0002x^2 + 0.0746x$, where y = global temperature change and x = global sea level rise, both from a 1970 baseline.

2.4. *Sectoral damage functions*

Methods and sources for regional and sectoral damage functions with respect to regional temperature, and in some case precipitation or sea level rise, were described in considerable detail in Neumann *et al.* (2020). Damages were scaled by factors appropriate for each sector (e.g., relevant population, infrastructure inventory, etc.) to facilitate a comparison of sensitivity across regions and to allow for independent forecasting of the relevant scalars over time.

Take the labor sector, for example. The damage function for labor (calibrated in terms of annual per capita wage loses) is

$$\text{DamageScaled}_{\text{labor}} = B0_{\text{labor}} + B1_{\text{labor}}(\text{DeltaT}) + B2_{\text{labor}}(\text{X\_ Region})$$
$$+ B3_{\text{labor}}(\text{X\_Region} \times \text{DeltaT}) + B4_{\text{labor}}(\text{X\_Era}).)$$

The function for roads (another sector, this time calibrated in terms of annual required repair costs per lane mile) is

$$\text{DamageScaled}_{\text{labor}} = B0_{\text{roads}} + B1_{\text{roads}}(\text{DeltaT}) + B2_{\text{roads}}(\text{X\_Region})$$
$$+ B3_{\text{roads}}(\text{X\_Regions} \times \text{DeltaT}) + B4_{\text{roads}}(\text{X\_Era}).)$$

In both examples, DeltaT is the change in temperature from the 1995 baseline (°C), X_Region is a region dummy variable, X_Region*DeltaT is the interaction of the region and temperature change variables, and X_era is the era dummy variable. Damage functions and regression results for all sectors are presented in Neumann *et al.* (2020) (see Table S4). An updated coastal property function representing the damages presented in this analysis is included in the Supplemental Materials of this paper, Table S4.

2.5. *Transient temperature pathways*

As noted above in Sec. 2.2, the transient calculations of GMST that were used to track damages across climate eras for discrete distributions of four alternative emissions scenarios were imported from Yohe (2017, Fig. 2). Following Neumann *et al.* (2020), estimated distributions of regional temperature change scenarios were calibrated against the 1986–2005 GMST average. They were all driven by the 12 GMST trajectories (5th, 50th, and 95th percentiles trajectories for each of the four emissions scenarios); and they were calibrated to four climate eras for the 21st century (2020–2039, 2040–2059, 2060–2079, and 2080–2099). The panels of Fig. 1 portray these results on maps which clearly identify the seven regions identified by the National Climate Assessment (NCA4, 2018); the graphs that are superimposed upon these maps also collectively display median, 5th, 95th percentile, and interquartile transient temperature trajectories for the designated region.

To establish context, it is important to check that these data (the products of a reduced form statistical approach anchored on projected data) are consistent with results from earlier process-based research on regional temperature projections for the end of the century. Those projections miss the intermediate climate eras, but the question here is "Do the temperature trajectories in Fig. 1 miss well-respected 2100 boundaries?" A quick review of assessments like NCA4 (2018) is reassuring in this regard. It shows, in Fig. 2 (Panel D), that process understanding of the climate supports the same pattern at the end of the century — higher anticipated warming along both SRES A2 and RCP 8.5 emissions scenarios (the "BAU" equivalents) in 2090 across the north and northeast parts of the country in comparison with the smaller temperature sensitivities is expected across the western, southwestern, and southeastern states.

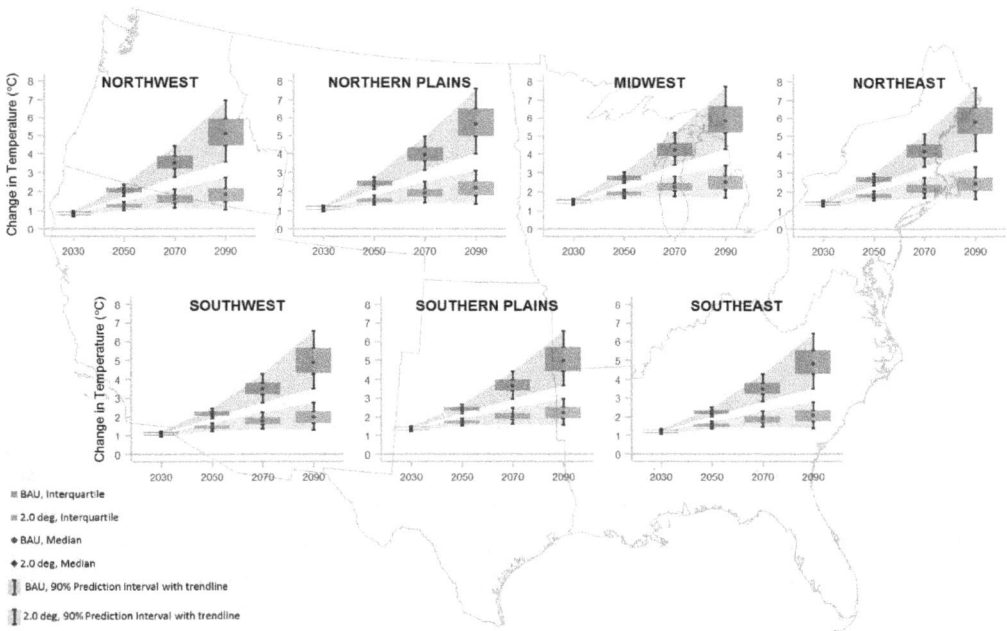

(a) Panel A: Regional temperature changes for Global "2.0" degree and "BAU" scenarios.

Source: Neumann *et al.* (2020).
Notes: As noted in the text, three of the four emissions scenarios keep transient temperatures below 1.5, 2, and 3 degrees Centigrade relative to preindustrial levels (0.85, 1.35 and 2.35 degrees Centigrade below the 1986 to 2005 climate era baseline). They are identified by their conventional names in the UNFCCC negotiations: that is, by "1.5", "2.0", "3.0", and "BAU" scenarios, respectively. The "Business as Usual" scenario is drawn from Fawcett *et al.* (2015); it is consistent with IPCC emissions pathways labeled RCP8.5 (IPCC, 2000) and SRES A2 (IPCC, 2008).

Figure 1. Regional transient temperature change trajectories. Regional maps of the transient regional trajectories are distributed for each NCA region across a map of the contiguous 48 states. Both panels depict the median, 5th, and 95th percentile estimates as well as inner quartile ranges for the midpoints of four consecutive 20-year climate eras from 2020 through 2100 compared to the 1986 to 2005 baseline era. Panel A contrasts the "BAU" and "2.0" degree scenarios; Panel B similarly contrasts "1.5" and "3.0" degree scenarios.

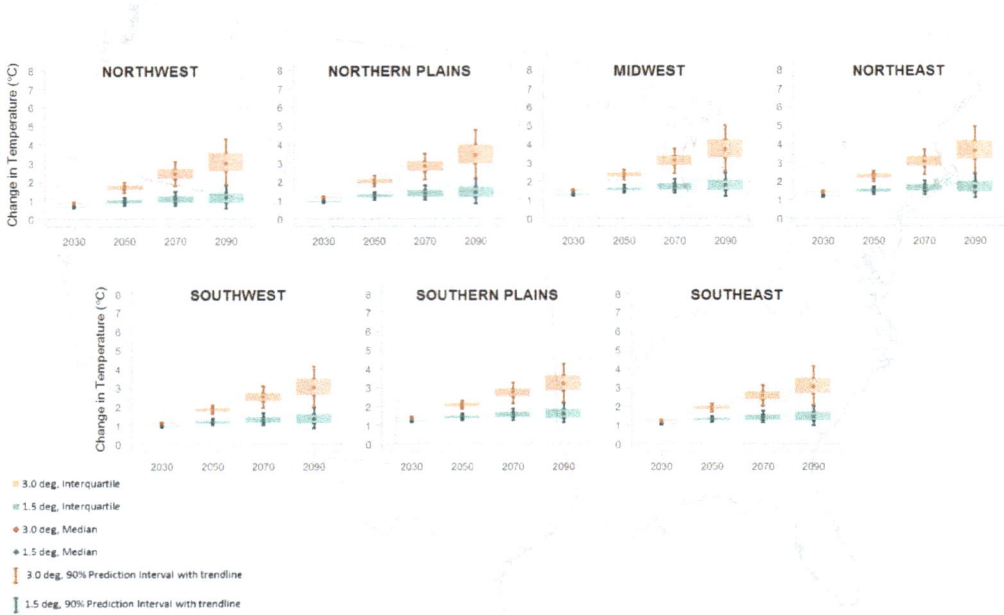

(b) Panel B: Regional temperature changes for the "1.5" and "3.0" degree scenarios.

Fig. 1. (*Continued*)

3. Mapping Damages Across the Continental US for Four Pathways

Distributions of transient sectoral damages were finally estimated for the four climate eras across these emissions-contingent regional distributions of temperature change on the basis of statistically estimated damage functions from Neumann *et al.* (2020). This text focuses primarily on damages to two illustrative sectors (labor and roads) so that both the analytical approach and its versatility are clear. Another 13 sectors are treated identically in Supplemental Material; some results from these other sectors are imported below into the discussion sections (see Table 1).

The damage estimate functional forms are reported in Sec. 2.4 of this paper but their content is most easily gleaned from another series of maps that highlight the variability of scaled damages across all seven regions. Figure 2 shows the results for labor and roads across all seven regions and for each of the four emissions-driven temperature change trajectories. These distributions depend, through *Monte-Carlo* simulation techniques, on GMST distributions due to uncertainty in the global climate system, and the skill of the empirical estimates. The two panels for each sector display the resulting temporal damage distributions calibrated for the four era benchmark years on a map of the contiguous 48 states; they present the 5th and 95th percentiles, interquartile ranges, and medians.

Higher damage sensitivity and regional temperature sensitivity are reflected for labor damages for the Midwest (MW), Southern Plains (SP), and Southeast (SE) in labor cost vulnerability over time followed closely by the high "BAU" emissions scenario by the Northern Plains (NP); Figure 1 shows background temperature sensitivity for these regions. The value of limiting temperatures to "1.5" degrees instead of "2.0" degrees seems to be modest, but adding another degree (i.e., moving from the "2.0" to "3.0" degree scenario) is relatively more costly, particularly in these same four regions. This is an important observation that is explored in more detail, below.

"BAU" is dramatically more damaging to roads relative to other futures that restrict temperatures. The three temperature limiting scenarios stack in order, but their ranges start negative and expand more noticeably only late in the century. In part because damages are primarily negative through 2050 along the two low emissions scenarios,

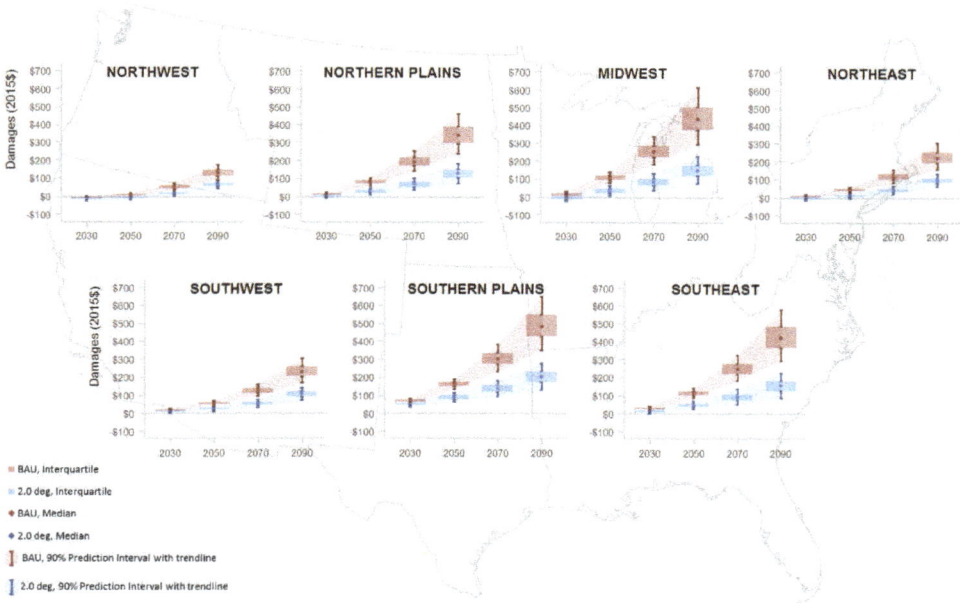

(a) Panel A: Labor Damages: Annual lost wages per capita by region and year for the "2.0" and "BAU" scenarios.

Note: As noted in the text, three of the four emissions scenarios keep transient temperatures below 1.5, 2, and 3 degrees Centigrade relative to preindustrial levels (0.85, 1.35 and 2.35 degrees Centigrade below the 1995 climate era baseline). They are identified by their conventional names in the UNFCCC negotiations: that is, by "1.5", "2.0", "3.0", and "BAU" scenarios, respectively. The "Business as Usual" scenario is drawn from Fawcett *et al.* (2015); it is consistent with IPCC emissions pathways labeled RCP8.5 (IPCC, 2000) and SRES A2 (IPCC, 2008).

Figure 2. Regional transient sectoral damage trajectories. For the labor and road sectors, regional damage trajectories (median, 5th and 95th percentiles, and the inner quartile range) are distributed across the contiguous 48 states for the four benchmark climate eras along all four of the emissions-driven GMT scenarios.

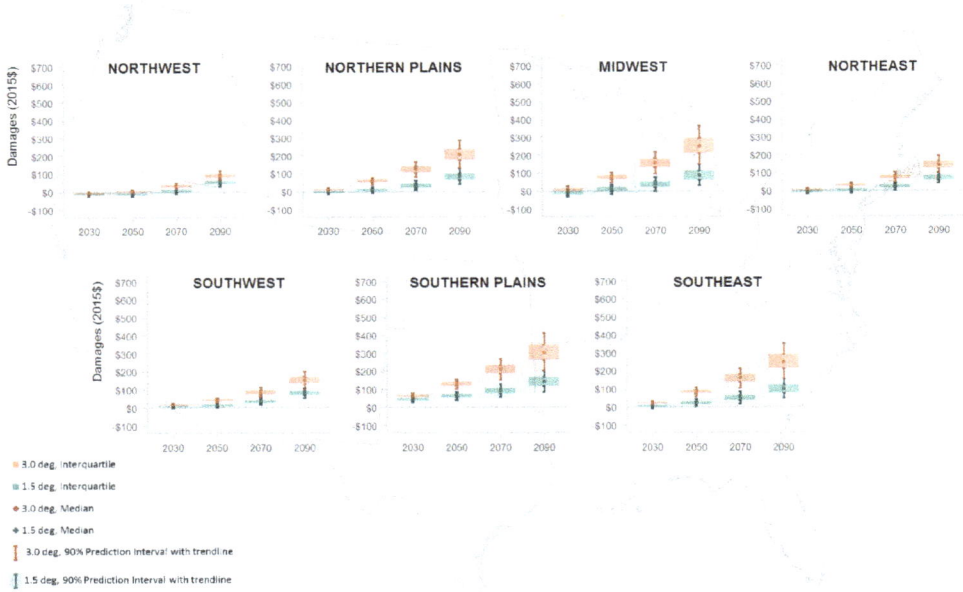

(b) Panel B: Labor Damages: Annual lost wages per capita by region and year for the "1.5" and "3.0" scenarios.

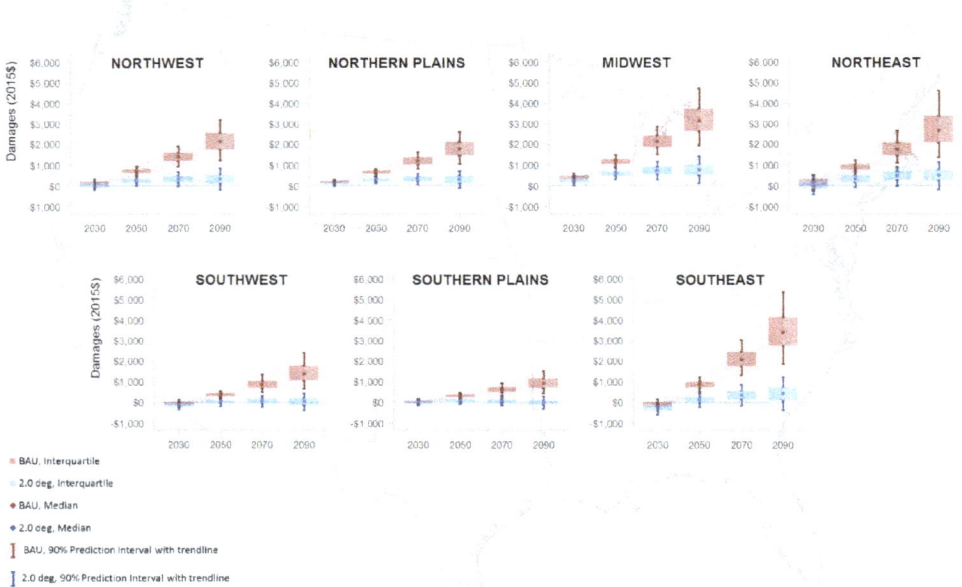

(c) Panel C: Road Damages: Annual required repair costs per lane mile per year by region and year for the "2.0" and "BAU" scenarios.

Fig. 2. (*Continued*)

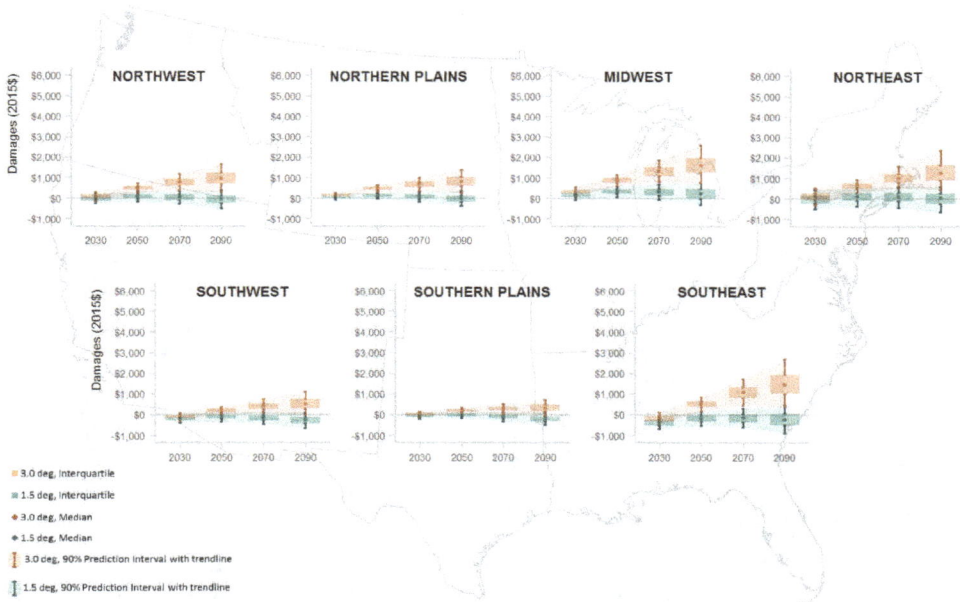

(d) Panel D: Road Damages: Annual required repair costs per lane mile per year by region and year for the "1.5" and "3.0" scenarios.

Fig. 2. (*Continued*)

there is, again, little value to these sectors in pushing an aspirational temperature maximum much below "2.0" degrees.

Table 1 summarizes a wider collection of findings across all 15 sectors in this analysis. That is, not only the two sectors covered in the text, but also the 13 sectors that occupy the Supplemental Material. These qualitative results were drawn from Fig. 2 and the various panels of Fig. S1, and they highlight the diversity of damage metrics that have been defined for individual sectors. Several persistent themes and conclusions emerge from a synthetic reading of the table. First, distributions for the BAU and the modest mitigation "3.0 Degree" scenarios generally bifurcate around midcentury for every region and most sectors. Thereafter, damages along the BAU trajectories are significantly higher than those calculated for the "3.0 Degree" target in a very precise sense — the 5th percentile pathway for the BAU case runs increasingly higher than the 95th percentile pathway for even modest mitigation. Put another way, moving future emissions below the BAU scenario can have *significant and growing* value in terms of avoided or postponed economic damages starting in the relatively near term.

Secondly, the Southern Plains (SP), Southeast (SE), and Northeast (NE) regions, followed by the Midwest (MW), seem to be most vulnerable to climate change risks for almost half of the modeled sectors. This means that, on a per unit basis, these

Table 1. Summary of Insights.

Sector / Damage Metric	Total Annual Damages BAU scenario, 2090 era		Scaled Annual Damages	
	National Damages (millions)	Regional Distribution	Damage Scalar	Scaled Damage Estimate Insights
Labor *Lost wages*	$150,000		*per capita*	• **"1.5"** and **"2.0"** negative early in century in **MW, NE,** and **NW** • Damages climb throughout century in all scenarios
Extreme Temperature *Lost value due to premature mortality*	$140,000		*per capita in selected cities*	• Damages in **"3.0"** at end of century are more than double those in **"2.0"** for **MW, NE,** and **SW** • Bifurcation between "2.0" and "BAU" scenarios in 2090 • **NW** damages more than an order of magnitude smaller than other regions by 2090 under **BAU** and **"3.0"**
Roads *Required repair costs*	$21,000		*per lane mile*	• **BAU** damages are more than double **"3.0"** at end of century in all regions • Damages increase over the century at a faster rate in BAU than other scenarios
Coastal Property *Property damage*	$20,000		*per dollar of coastal property value*	• Only **BAU** damages increase in all regions across the century • **SP** has largest damage per coastal property value, particularly early in the century where **SP** damages are about 2x next highest region
Electricity Supply and Demand *Change in system cost*	$8,000		*per capita*	• **NW**: region with largest damages ($32 per capita) in 2090 under **BAU** and the largest increase between 2050 and 2090 in all scenarios • **BAU** damages are nearly double those of "**3.0**" in all regions in 2090 • Average **BAU** damages more than triple across century; all other scenarios have increases of less than double
Urban Drainage *Adaptive costs*	$5,300		*per square mile of urban area in selected cities*	• Damages remain relatively flat over time in all scenarios. Damages decrease for the majority of regions between 2030 and 2090 in "1.5" and "2.0" scenarios. • No bifurcation between scenarios

Table 1. (*Continued*)

Sector / *Damage Metric*	Total Annual Damages *BAU scenario, 2090 era* National Damages (millions)	Total Annual Damages *BAU scenario, 2090 era* Regional Distribution	*Damage Scalar*	Scaled Annual Damages Scaled Damage Estimate Insights
Rail *Delay and temperature sensor installation costs*	$5,300	(regional map)	*per rail mile*	• 2070 damages lower than 2050 damages for most regions and scenarios • All scenarios negative early in century except MW • Trends across scenarios are similar; magnitudes differ, increasing with warmer trajectories
Water Quality *Water quality improvement costs*	$4,800	(regional map)	*per capita*	• Little variation in increasing damage trajectories across regions
Freshwater Fish *Lost recreation value*	$3,800	(regional maps)	*per capita*	• Damages generally decrease across century under "**1.5**" and "**2.0**" • Negative damages (i.e. benefits) in NW for all scenarios across the century • Negative damages (i.e. benefits) in 2070 and 2090 in all regions but NE and SP under "**1.5**"
West Nile Virus *Lost value due to premature mortality*	$3,200	(regional maps)	*per capita*	• NP damages at least 3x larger than other regions in "**3.0**" and "**BAU**" • Damages increase across century for all scenarios in all regions
Winter Recreation *Lost recreation value*	$2,000	(regional maps)	*per capita*	• Negative damages (i.e. benefits) for all regions under "**1.5**" in 2090 • All scenarios have negative damages in some regions in 2090 except **BAU**
Bridges *Required repair costs*	$910	(regional map)	*per susceptible bridge for reactive repairs*	• Damages decrease across century in all regions and scenarios • Very little bifurcation between scenarios, even in 2090

Table 1. (*Continued*)

Sector / *Damage Metric*	Total Annual Damages BAU *scenario, 2090 era*		Damage Scalar	Scaled Annual Damages / Scaled Damage Estimate Insights
	National Damages (millions)	Regional Distribution		
Municipal and Industrial Water Supply / *Value of unmet water demand*	$340		*per capita*	• Bifurcation between scenarios in **MW** and **SP** only • **SE** and **NW** have negative damages (i.e. benefits) in the first half of the century under all scenarios
Harmful Algal Bloom / *Lost recreational value*	$240		*per capita*	• All scenarios negative in **MW** and **NW** early in century • **SW** damages smaller than **SE** by an order of magnitude under **BAU** in 2090
Aeroallergens / *Emergency department visitation costs*	$1.7		*per capita in selected cities*	• All scenarios negative in **SE** early in century; "**1.5**" and "**2.0**" negative in 2090. • Damages increase by at least 25% between "**1.5**" and "**2.0**"; damages more than double between "**2.0**" and "**3.0**" in 2090 • Damages peak in mid-century for "**1.5**" and "**2.0**"

Source: Figure 2 and Fig. S2 in the Supplemental Material. Regions are identified as follows: Northwest = NW, Northern Plains = NP, Midwest = MW, Northeast = NE, Southwest = SW, Southern Plains = SP, and Southeast = SE. "Bifurcation", noted in the estimate insights, is defined as no overlap in the 90% prediction intervals (5th to 95th percentiles). In Regional Distribution maps, gradient color ranges from highest benefits by sector in dark blue to highest damages by sector in dark red. Regions in white were not modeled for that particular sector.

Note: Regions were not modeled for one of two reasons: 1. No damages are expected to occur in that region/sector combination (Coastal and Winter Recreation); 2. The technique has not yet been applied to resources in that region (Extreme Temperature and Aeroallergens). The latter reason is being addressed in subsequent work.

Table 2. Total damages by sector for the contiguous 48 states.

| Era | Global Temperature Scenario | | | |
	"1.5"	"2.0"	"3.0"	"BAU"
Panel A: U.S. total annual damages in labor (million $).				
2030	$1,500	$3,700	$6,300	$7,600
	(−$3,500; $6,500)	(−$1,100; $8,500)	($1,600; $11,000)	($3,100; $12,000)
2050	$6,800	$14,000	$25,000	$34,000
	(−$410; $14,000)	($7,000; $21,000)	($17,000; $32,000)	($26,000; $42,000)
2070	$17,000	$31,000	$53,000	$81,000
	($6,200; $29,000)	($18,000; $46,000)	($35,000; $70,000)	($60,000; $110,000)
2090	$39,000	$59,000	$91,000	$150,000
	($22,000; $60,000)	($36,000; $83,000)	($59,000; $130,000)	($100,000; $200,000)
Panel B: U.S. total annual damages in roads (million $).				
2030	−$710	−$180	$430	$760
	(−$2,900; $1,400)	(−$2,200; $1,900)	(−$1,500; $2,400)	(−$1,200; $2,700)
2050	$460	$2,100	$4,500	$6,400
	(−$1,900; $2,700)	(−$160; $4,300)	($2,300; $6,600)	($4,300; $8,700)
2070	$190	$3,100	$7,400	$13,000
	(−$2,900; $3,200)	(−$180; $6,300)	($3,600; $11,000)	($8,600; $19,000)
2090	$530	$3,300	$9,400	$21,000
	(−$4,800; $3,700)	(−$1,700; $8,100)	($3,000; $17,000)	($12,000; $33,000)
Panel C: U.S. total annual damages for mortality from extreme heat (million $) The underlying extreme heat damage models only produced results for two eras: 2050 and 2090.				
2050	$250	$8,600	$21,000	$31,000
	(−$12,000; $12,000)	(−$3,400; $20,000)	($9,000; $32,000)	($19,000; $42,000)
2090	$12,000	$34,000	$70,000	$140,000
	(−$12,000; $36,000)	($5,500; $62,000)	($33,000; $110,000)	($86,000; $200,000)
Panel D: U.S. total annual damages to coastal property (million $)				
2030	$4,600	$4,700	$4,900	$5,000
	($4,000; $5,200)	($4,100; $5,400)	($4,300; $5,600)	($4,400; $5,700)
2050	$5,000	$5,500	$6,300	$7,200
	($4,300; $5,800)	($4,700; $6,400)	($5,400; $7,400)	($6,100; $8,400)
2070	$5,300	$6,300	$8,300	$12,000
	($4,300; $6,500)	($5,100; $7,900)	($6,400; $11,000)	($9,000; $17,000)
2090	$5,000	$6,300	$9,400	$20,000
	($3,800; $6,500)	($4,600; $8,700)	($6,200; $15,000)	($11,000; $45,000)

Notes: Total costs (in millions of dollars) across the contiguous 48 states for the four 20-year climate eras between 2020 and 2100 (labeled by their midpoints) along all four of the emissions-driven GMT scenarios. Values provided are median estimates and 5th and 95th percentile range estimates. Three of the four emissions scenarios keep transient temperatures below 1.5°C, 2°C, and 3°C relative to preindustrial levels (0.85°C, 1.35°C, and 2.35°C below the 1995 climate era baseline). They are identified by their conventional names in the UNFCCC negotiations: that is, by "1.5 Degrees", "2.0 Degrees", "3.0 Degrees", and "BAU" scenarios, respectively. The "Business as Usual" scenario is drawn from Fawcett *et al.* (2015); it is consistent with IPCC emissions pathways labeled RCP8.5 (IPCC, 2000) and SRES A2 (IPCC, 2008).

regions see the largest damages relative to their baseline values of economic significance. The sources of these vulnerabilities vary in importance across the regions, but the ranking is rigorously substantiated.

In addition, some damages for some sectors are actually negative (i.e., they are benefits), especially in the early climate era or two; but most damages (though not all) increase at an increasing rate over time as the country moves from earlier climate eras to the later ones. This means that even negative sectoral damages can and usually do turn positive later in the century.

4. Total Sectoral Damages

Table 2 records the results of the national level calculations for the labor and roads sectors, as well as extreme temperature mortality and coastal property value (two of the 13 sectors included in the Supplemental Materials). Table S2 displays the same information for the remaining 11. While it is not appropriate to claim that the sum of estimated damages across the small number of sectors in the Supplemental Materials would be a meaningful reflection of total economic risk for the United States, it is appropriate to add damages for each sector across the seven regions to estimate 15 sectoral damage aggregates.

In absolute terms, for the contiguous US, total labor damages (Panel A) are nearly an order of magnitude larger than total road sector damages (Panel B) for all three reported statistics and all four emissions trajectories by, at the latest, 2090. Indeed, due to reduced damages associated with a lessening of freeze-thaw damage, total road repair costs are actually negative early along the lower temperature constrained emissions scenarios in the first climate era. Extreme temperature mortality damages are very similar in pattern and magnitude to labor damages.

It is, again, interesting to place these totals in the context of earlier work. Hsiang *et al.* (2017), for example, provide econometrically-based estimates for a variety of sectors for the United States from data drawn in large measure from the county level. They find that the poorest counties could see up to 20% losses in income by the end of the century in a RCP8.5 (BAU) scenario. For low and high risk labor combined, they reported a median aggregate estimate of roughly $95 billion in 2090 surrounded by a 5th percentile estimate of $74 billion and a 95th percentile estimate of $114 billion. Our reduced form regional projection approach worked with historical regional antecedents to produce higher labor damages for the same time frame: a median of $150 billion surrounded by $100 and $200 billion for 5th and 95th percentiles, respectively. Perhaps the difference can be explained by differences in the scaling of future wage increases; our labor pool was more inclusive. Or perhaps our aggregation missed ameliorating micro-scale variation. Or perhaps our higher estimates were caused by our "BAU" scenario's displaying higher emissions mid-century on the way to more than 100 GtCO2 by 2085. In any case, they are in the same order of magnitude. We think that our results are qualitatively significant and report that, by the end of the century, annual wages per capita will be much reduced. It should be noted that a regional distribution along our "BAU" scenario suggests that the Northeast, the Southeast, and the Midwest are the most vulnerable regions for our labor metric — an observation that coincides well with Hsiang and colleagues even though they included the Texas portion of the Southwest in that category.

Panel C of Table 2 reports total annual damage estimates for extreme temperature mortality. They are contingent on the "value of statistical life", and they climb to $31 billion (plus or minus $12 billion) in 2050 and $140 billion (plus or minus $55 to $60 billion) by the end of the century in the "BAU" scenario. The variance is clearly enormous, but even the 5th percentile dominates other sectors shown only in Supplemental Materials. Net extreme temperature damages also start in the negative range for low emissions scenarios for both 2050 and 2090, a result of larger reductions in cold-related mortality compared to increases in heat-related deaths. Mortality from heat events is clearly a sector where reducing emissions as quickly and as effectively as possible (to get back to 1.5°C or 2.0°C) could pay dividends through mid-century as well as into the 2090 climate era compared to the "3.0 Degree" target trajectories.

Estimated annual damages to coastal property, displayed in Panel D of Table 2, speak to a sector that has a been a poster child not only for extreme events caused by dangerous hurricanes and cyclones, but also more routine and less severe coastal events whose damages can climb with sea level rise. They are reported in Panels O and P of Fig. S1 as fractions of the total value of coastal property in each region that are generally lower than 0.005, or $5 per $1000 of coastal property. According to this metric, the Southern Plains (i.e., the Texas coastline) shows the largest vulnerability of the five relevant regions. The Southern Plains distributions also display, by far, the largest variation around the median trajectories for all emissions scenarios. The absolute coastal damage estimates are comparable with those produced for the roads sector (except that they are not negative in the 2030 climate era) due to the large total value of coastal property across all coastlines. The Southeast region makes up about 75% of all absolute damages due to its vulnerability and large stock of coastal properties.

It is important to remember, when reading Table 2, that the sectoral results typically reflect "no adaptation" vulnerability data. Factoring empirical representations of adaptation into their calculations could reduce these estimates by as much as 85% (see Neumann *et al.*, 2020), but Martinich and Crimmins (2019) argue that this reduction can be less (or a little more) under certain assumptions about existing levels of adaptation investments (the projection boundary condition for total value). The effect of adaptation is further explored in Sec. 6. Damages in 2030 are about $5 billion across all emission scenarios. In the higher emissions scenarios ("3 Degree" and "BAU"), the estimates grow measurably through the end of the century. They sum to $20 billion (minus $9 billion but plus $25 billion) for the 2090 climate era along BAU, for example — up from $5 billion (plus or minus $600 million) for the 2030 climate era. Interestingly, the "1.5 Degree" future sees damages rise and fall slowly within the boundaries of the 5th to 95th percentile cone a median of roughly $5 billion in the 2030 climate era. The other two mitigation scenarios show 50% to 100% growth across the four climate eras. Given the experiences in the Southeast and Southern Plains regions in 2017, 2018, and 2019,[4] however, it is a stretch to think that these estimates credibly account for risks associated with extreme events that are now projected with greater confidence to increase in intensity; see, for example, very recent corroborating work drawn from data spanning the past four decades for storms that can threaten or assault the Southern Plains and Southeast coastlines in Kossin *et al.* (2020).

5. Comparing "1.5 Degree" Futures with "2.0 Degree" Futures

IPCC (2018b) is an official response to an invitation "to provide a Special Report in 2018 on the impacts of global warming of 1.5°C above pre-industrial levels and related global greenhouse gas emission pathways".[5] The IPCC decided to accept the invitation in April of 2016 by preparing a special report "in the context of strengthening the global response to the threat of climate change, sustainable development, and efforts to eradicate poverty". For present purposes, two conclusions are particularly germane. Firstly, from two pages of headlines, "Climate-related risks for natural and human systems are higher for global warming of 1.5°C than at present, but lower than at 2°C (*high confidence*). These risks depend on the magnitude and rate of warming, geographic location, levels of development, and vulnerability, and on the choices and implementation of adaptation and mitigation options (*high confidence*)". Later, conclusion B.5.5 reports that "Risks to global aggregated economic growth due to climate change impacts are projected to be lower at 1.5°C than at 2°C by the end of this century (*medium confidence*). This excludes the costs of mitigation, adaptation investments, and the benefits of adaptation. Countries in the tropics and Southern Hemisphere subtropics are projected to experience the largest impacts on economic growth due to climate change should global warming increase from 1.5°C to 2°C (*medium confidence*)".

Interestingly, the economic conclusions were advanced with only "medium" confidence. To truly understand what that means and why, it is important to review the uncertainty (language) guidance provided to IPCC authors of all subsequent assessments by Mastrandrea *et al.* (2010). There, it is clear that this designation signals concerns about the quality and quantity of underlying data and/or the degree of agreement over the understanding of underlying processes across a diversity of different sectors. These concerns are especially important when the risk metric of choice

---

[4]Total damage estimates for 2017, Hurricanes Irma (SP), Maria (SE), and Harvey (SP) — $77, $92, and $128 billion, respectively, total $297 billion; for 2018, Hurricanes Florence (SE) and Michael (SP) — $24 and $25 billion, respectively, total $49 billion; for 2019, Irma and Dorian — $4 and $10 billion, respectively, total a mere $14 billion. Source: https://www.ncdc.noaa.gov/billions/time-series.

[5]The invitation was Decision 1/CP.21 (paragraph 21) in the Report of the 21st Conference of the Parties of the United Nations Framework Convention on Climate Change contained in the Decision of the 21st Conference of Parties of the United Nations Framework Convention on Climate Change to adopt the Paris Agreement.

is an aggregate like gross domestic product. In addition, a majority of the cited economic literature was tied only to changes in temperature. The available estimates of economic costs were therefore usually independent of time and disconnected from the socio-economic content of any mitigation pathway. Here, we have already warned strongly that summing across an incomplete list of sectors would be misleading, and we argue even more strongly that our tying estimates to temporal and mitigation-specific scenarios for various regions and sectors in the United States begins to fill a critical informational gap.

More specifically, as shown in Fig. 3, the economic cost distributions of damages along the "1.5 Degree" and "2.0 Degree" scenarios persistently overlap throughout the 21st century for nearly all sectors and regions. That is to say, bifurcations between the 5th percentiles for the 2°C temperature target and 95th percentiles for the 1.5°C target of the do not appear by 2090 — at least for the 12 most vulnerable sectors that are studied here. Indeed, for all sectors except bridges and roads, the median estimates for "1.5 Degree" and "2.0 Degrees" *both* lie inside the overlapping inner-quartile ranges of the alternative target. Take, for example, coastal damages. The value of achieving a "1.5 Degree" maximum relative to a "2.0 Degree" maximum is not very large (differences in medians less than $500 million through the 2040 era and $1.3 billion by the end of the century. It follows that striving to return to a "1.5 Degree" maximum relative

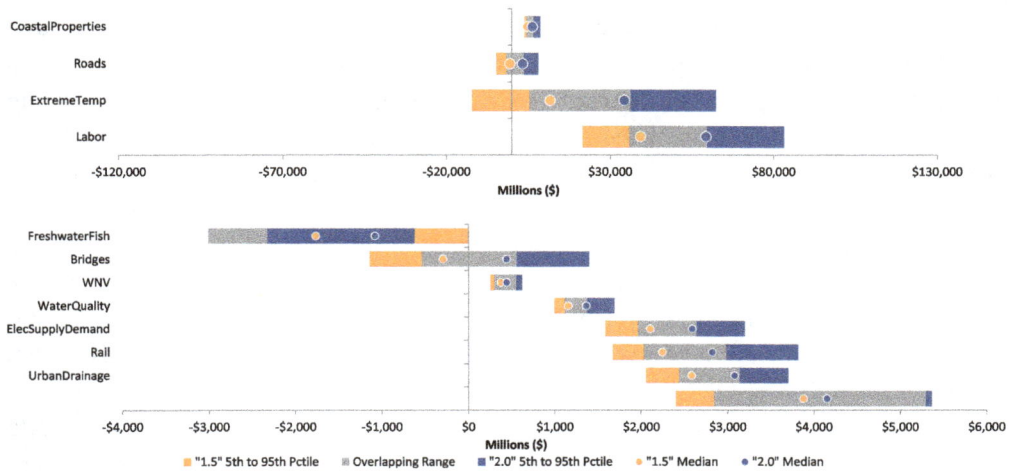

Figure 3. "1.5" versus "2.0" damages by sector in 2090. 5th to 95th percentile range of total costs (in millions of dollars) across the contiguous 48 states by sector for the "1.5" degree (in orange) and "2.0" degree scenarios (in blue). Overlapping ranges are shown in gray. Median estimates for each sector and scenarios are shown as dots in the color representing the scenario. Note the differences in scales between the two panels due to large variation in magnitude across sectors. Three sectors (Aeroallergens, Harmful Algal Blooms, and Municipal and Industrial Water Supply) are not shown, as the magnitude of damages is negligible compared to other sectors.

to a "2.0 Degree" maximum would produce little statistically justified value in terms of avoided economic damage across the studied sectors.

6. Concluding Remarks

The text and associated Supplemental Materials contribute internally consistent and therefore entirely comparable regional, temporal, and sectoral risk profiles to a growing literature on regional vulnerability to climate change. A large collection of maps populated with graphs of *Monte-Carlo* simulation results support a communication device in this regard — a convenient visual that we hope will make comparative results tractable and credible and resource allocation decisions more transparent.

Since responding to climate change is a risk-management problem, it is important to note that these results speak to both sides of the risk calculation. They characterize likelihood distributions along four alternative emissions futures (thereby reflecting the mitigation side context); and they characterize consequences along these transient trajectories (which can thereby inform planning for the iterative adaptation side). Looking across the abundance of sectors that are potentially vulnerable to some of the manifestations of climate change, the maps therefore hold the potential of providing comparative information about the magnitude, timing, and regional location of relative risks. This is exactly the information that planners who work to protect property and public welfare by allocating scarce resources across competing venues need to have at their disposal — information about relative vulnerabilities across time and space and contingent on future emissions and future mitigation. It is also the type of information that integrated assessment researchers need to calibrate and update their modeling efforts — scholars who are exemplified by Professor Nordhaus who created and exercised the Dynamic Integrated Climate-Economy and Regional Integrated Climate-Economy models (Nordhaus, 1994, 2008; Nordhaus and Boyer, 2000). Both models were drawn from the same conceptual foundations, but the former is global and the latter is disaggregated across multiple economically differentiated regions. Both are widely respected, both have been applied frequently by other researchers from around the world, and both ultimately provided the context and background for his Nobel citation and his Nobel Lecture (Nordhaus, 2018).

The text focused attention on two sectors by reporting lost annual wages per capita and the annual required repair cost per lane mile and provided additional commentary on a third sector, coastal property, and some discussion of extremes. A diversity of results emerged even from this limited coverage. Annual risks from lost wages dominated annual road damages in every region, especially for low emissions trajectories where road damages were negative in nearly every region. Lowering maximum global mean temperatures from "2.0 Degree" to "1.5 Degree" maxima did not significantly lower risk across the country, but reducing "BAU" emissions to levels that would sustain a "2.0" degree maximum (or even a "3.0" maximum) did so, especially for labor in the Midwest, Northeast, Southern Plains, and the Southeast. For roads, holding temperature increases below a "2.0" degree maximum produces advantages by reducing freeze/thaw events in the early part of the century; total road damages become positive, though, above a "3.0" degree maximum emissions trajectory around 2030 even in the southern regions, as direct temperature and precipitation effects on pavement integrity overtake the beneficial effects of reductions or northward migration in freeze-thaw phenomena. Again, the Midwest, Northeast, and Southeast dominated risk as emissions grow toward the "BAU" limiting case.

Supplemental Materials report the results of replicating the analysis for 13 other sectors that are, to varying degrees, important across the contiguous 48 states. Mortality risk from extreme heat dominated those estimates for all climate eras along all emissions scenarios.

The value of limiting temperature below a "1.5 Degrees" maximum compared to "2.0 Degree" maximum is persistently small across sectors and regions. By way of contrast, statistically significant bifurcations of damage estimates relative to BAU emissions generally appear mid-century or later for most regions and most sectors compared to even the "3.0" temperature constraining trajectories. Taken as a whole, the Midwest, Northeast, Southeast, and Southern Plains regions seem to be more vulnerable along high emissions scenarios for nearly a majority of their important sectors in the relatively near term.

Finally, the damage estimates generally reflect explicit "no adaptation" scenarios, though in some cases where econometric estimates underlie the sectoral results (e.g., labor), damage estimates reflect historical adaptation responses. All bets are off looking forward, there, given the extreme disruptions of economic responses to the COVID-19 virus.

It is currently possible to look at the economics of adaptation for a small subset of sectors, such as coastal property and the infrastructure sectors (see Neumann *et al.*, 2020; which examines a range of adaptation responses), but data to calibrate these scenarios for other econometrically estimated sectoral damages do not exist because we have little history of adaptation to temperature changes as large as those we examine here. Some caution should be exercised, however, in interpreting the likelihood of full adoption of even cost-effective adaptation, in light of current history of uneven and/or delayed uptake of adaptation in the U.S. — see for example Lorie *et al.* (2020) and Carey (2020). As outlined in Chambwera *et al.* (2014), while the economics of adaptation may identify numerous cost-effective options for lessening the damages of climate change, practical issues of financial limitations, site-specific technical and physical limits, multiple objectives, and other implementation constraints limit the timely or even eventual adoption of these measures. Suboptimal adaptation is therefore implied to be the norm. Nonetheless, envisioning a series of these calculations with incremental inclusion of suggested temporal, sectoral, and regional allocations of resources to adaptation investments could be informative in implementing the adjective "iterative" in the IPCC climate risk management conclusion with which we began.

Acknowledgments

This research was partially funded by the U.S. Environmental Protection Agency under contract EP-D-14-031. The views expressed in this document are those of the authors and do not necessarily reflect those of the U.S. Environmental Protection Agency.

References

Auffhammer, M (2018). Quantifying economic damages from climate change. *Journal of Economic Perspectives*, 32, 33–52.

Carey, L (2020). Donald Trump is right. We need 'BIG & BOLD' infrastructure spending. Center for Strategic and International Studies. April 6. Available at: https://www.csis.org/analysis/donald-trump-right-we-need-big-bold-infrastructure-spending.

Chambwera, M, G Heal, C Dubeux, S Hallegatte, L Leclerc, A Markandya, B McCarl, R Mechler and J Neumann (2014). Economics of adaptation. In *Climate Change 2014: Impacts, Adaptation, and Vulnerability. Part A: Global and Sectoral Aspects. Contribution of Working Group II to the Fifth Assessment Report of the Intergovernmental Panel on Climate Change*, Chap. 17, CB Field, VR Barros, DJ Dokken, KJ Mach, MD Mastrandrea, TE Bilir, M Chatterjee, KL Ebi, YO Estrada, RC Genova, B Girma, ES Kissel, AN Levy, S MacCracken, PR Mastrandrea and LL White (eds.), pp. 945–977. Cambridge, United Kingdom and New York, NY, USA: Cambridge University Press.

Collins, WJ, N Bellouin, M Doutriaux-Boucher, N Gedney, P Halloran, T Hinton, J Hughes, CD Jones, M Joshi, S Liddicoat, G Martin, F O'Connor, J Rae, C Senior, S Sitch, I Totterdell, A Wiltshire and S Woodward (2011). Development and evaluation of an Earth system model–HadGEM2. *Geoscience Model Development*, 4, 1051–1075.

Donner, LJ *et al.*, (2011). The dynamical core, physical parameterizations, and basic simulation characteristics of the atmospheric component of the GFDL global coupled model CM3. *Journal of Climate*, 24, 3484–3519, https://doi.org/10.1175/2011JCLI3955.1.

Fawcett, A, G Iyer, L Clarke, J Edmonds, N Hultman, H McJeon, J Rogelj, R Schuler, J Alsalam, G Asrar, J Creason, M Jeong, J McFarland, A Mundra and W Shi (2015). Can Paris pledges avert severe climate change? *Science*, 350, 1168–1169.

Gent, PR, G Danabasoglu, LJ Donner, MM Holland, E Hunke, S Jayne, D Lawrence, RB Neale, PJ Rasch, M Vertenstein and PH Worley (2011). The community climate system model version 4. *Journal of Climate*, 24, 4973–4991.

Hsiang, S, R Kopp, A Jina, J Rising, M Delgado, S Mohan, D Rasmussen, R Muir-Wood, P Wilson, M Oppenheimer, L Larsen and T Houser (2017). Estimating economic damage from climate change in the United States. *Science*, 356, 1362–1369, https://doi.org/10.1126/science.aal4369.

Intergovernmental Panel on Climate Change (IPCC) (2000). *Special Report on Emissions Scenarios: A special report of Working Group III of the Intergovernmental Panel on Climate Change*. Cambridge, United Kingdom and New York, NY, USA: Cambridge University Press, www.IPCC.ch.

Intergovernmental Panel on Climate Change (IPCC) (2007). *Synthesis Report of the Fourth Assessment Report*. Cambridge, United Kingdom and New York, NY, USA: Cambridge University Press, www.IPCC.ch.

Intergovernmental Panel on Climate Change (IPCC) (2008). *Towards New Scenarios for Analysis of Emissions, Climate Change, Impacts, and Response Strategies*, p. 132. Geneva: IPCC, www.IPCC.ch.

Intergovernmental Panel on Climate Change (IPCC) (2014). *Synthesis Report of the Fifth Assessment Report.* Cambridge, United Kingdom and New York, NY, USA: Cambridge University Press, www.IPCC.ch.

Intergovernmental Panel on Climate Change (IPCC) (2018a). Special report on Global Warming of 1.5°C. World Meteorological Organization, Geneva, Switzerland, www.IPCC.ch.

Intergovernmental Panel on Climate Change (IPCC) (2018b). Summary for policymakers. In *Global Warming of 1.5°C. An IPCC Special Report on the Impacts of Global Warming of 1.5°C above Pre-industrial Levels and Related Global Greenhouse Gas Emission Pathways, in the Context of Strengthening the Global Response to the Threat of Climate Change, Sustainable Development, and Efforts to Eradicate Poverty,* V Masson-Delmotte, P Zhai, H-O Pörtner, D Roberts, J Skea, PR Shukla, A Pirani, W Moufouma-Okia, C Péan, R Pidcock, S Connors, JBR Matthews, Y Chen, X Zhou, MI Gomis, E Lonnoy, T Maycock, M Tignor and T Waterfield (eds.), p. 32. Geneva, Switzerland: World Meteorological Organization.

Kopp, RE, RM Horton, CM Little, JX Mitrovica, M Oppenheimer, DJ Rasmussen, BH Strauss and C Tebaldi (2014). Probabilistic 21st and 22nd century sea-level projections at a global network of tide-gauge sites. *Earth's Future*, 2(8), 383–406, https://doi.org/10.1002/2014EF000239.

Kossin, J, R Knapp, T Olander and C Velden (2020). Global increase in major tropical cyclone exceedance probability over the past four decades. *Proceedings of the National Academies of Science*, 117(22), 11975–11980, https://www.pnas.org/content/117/22/11975.

Lorie, M, JE Neumann, MC Sarofim, R Jones, RM Horton, RE Kopp, C Fant, C Wobus, J Martinich, M O'Grady, LE Gentile (2020). Modeling coastal flood risk and adaptation response under future climate conditions. *Climate Risk Management*, 29, 100233, https://doi.org/10.1016/j.crm.2020.100233.

Martinich, J and A Crimmins (2019). Climate damages and adaptation potential across diverse sectors of the United States. *Nature Climate Change*, 9, 397–404.

Mastrandrea, MD, CB Field, TF Stocker, O Edenhofer, KL Ebi, DJ Frame, H Held, E Kriegler, KJ Mach, PR Matschoss, G-K Plattner, GW Yohe and FW Zwiers (2010). Guidance note for lead authors of the IPCC Fifth Assessment Report on consistent treatment of uncertainties. Intergovernmental Panel on Climate Change (IPCC). Available at http://www.ipcc.ch.

National Research Council (NRC) (2010). *Climate Stabilization Targets—Emissions, Concentrations, and Impacts over Decades to Millennia.* Washington DC, USA: National Academies Press.

Neumann, JE, J Willwerth, J Martinich, J McFarland, MC Sarofim and G Yohe (2020). Climate damage functions for estimating the economic impacts of climate change in the United States. *Review of Environmental Economics and Policy*, 14(1), 25–43.

Neumann, JE, P Chinowsky, J Helman, M Black, C Fant, K Strzepek and J Martinich (2020). Climate effects on US infrastructure: The economics of adaptation for rail, roads, and coastal development. Working Paper available at www.indecon.com/projects/benefits-of-global-action-on-climate-change/.

Nordhaus, W (1994). *Managing the Global Commons: The Economics of Climate Change.* Cambridge, MA: MIT Press.

Nordhaus, W and J Boyer (2000). *Warming the World: Economic Modeling of Global Warming.* Cambridge, MA: MIT Press.

Nordhaus, W (2008). *A Question of Balance: Weighing the Options on Global Warming Policies.* New Haven, CT, USA: Yale University Press.

Nordhaus, W (2018). Climate change: The ultimate challenge for economics. *Prize Lecture*, NobelPrize.org, Nobel Media AB 2020. 24 June. Available at https://www.nobelprize.org/prizes/economic-sciences/2018/nordhaus/lecture/.

Sarofim, MC, J Martinich, JE Neumann, J Willwerth, Z Kerrich, M Kolian, C Fant and C Hartin (2020). A temperature-binning approach for multi-sector climate impact analysis. Working Paper available at www.indecon.com/projects/benefits-of-global-action-on-climate-change/

Schmidt, GA, R Ruedy, JE Hansen, I Aleinnov, N Bell, M Bauer, S Bauer, B Cairns, V Canuto, Y Cheng and A Del Genio (2006). Present-day atmospheric simulations using GISS ModelE: Comparison to in situ, satellite, and reanalysis data. *Journal of Climate*, 19, 153–192.

Sweet, WV, RE Kopp, CP Weaver, J Obeysekera, RM Horton, ER Thieler and C Zervas (2017). Global and regional sea level rise scenarios for the United States. NOAA Technical Report NOS CO-OPS 083. NOAA/NOS Center for Operational Oceanographic Products and Services.

United States Global Change Research Program (NCA3) (2014). Third National Climate Assessment, www.nca2014.globalchange.gov.

United States Global Change Research Program (NCA4) (2018). *Fourth National Climate Assessment*, www.nca2018.globalchange.gov.

U.S. Environmental Protection Agency (USEPA) (2017). Multi-model framework for quantitative sectoral impacts analysis: A technical report for the Fourth National Climate Assessment. EPA 430-R-17-001. Available at https://cfpub.epa.gov/si/si_public_record_report.cfm?Lab=OAP&dirEntryId=335095.

von Salzen, K, JF Scinocca, NA McFarlane, J Li, JN Cole, D Plummer, D Verseghy, MC Reader, X Ma, M Lazare and L Solheim (2013). The Canadian fourth generation atmospheric global climate model (CanAM4). Part I: Representation of physical processes. *Atmosphere-Ocean*, 51, 104–125.

Watanabe, M, T Suzuki, R O'ishi, Y Komuro, S Watanabe, S Emori, T Takemura, M Chikira, T Ogura, M Sekiguchi and K Takata (2010). Improved climate simulation by MIROC5: Mean states, variability, and climate sensitivity. *Journal of Climate*, 23, 6312–6335.

Yohe, G (2017). Characterizing transient temperature trajectories for assessing the value of achieving alternative temperature targets. *Climatic Change*, 145, 469–479. www.gyohe.faculty.wesleyan.edu.

CHAPTER 3

THE AUDACIOUS DR. NORDHAUS

LAWRENCE H. GOULDER

Department of Economics, Stanford University
Research Associate, National Bureau of Economic Research
University Fellow, Resources for the Future
goulder@stanford.edu

Keywords: Audacity in research; professional courage; economics of climate change; DICE model, price of light.

Much has been and will continue to be said about the extraordinarily important intellectual insights that Bill Nordhaus has contributed to several fields of economics, including, but not limited to, environmental economics. Happy to join the enormous chorus of appreciative fans.

But there's another dimension of Bill's work that deserves acknowledgment. Beyond the intellectual insights, something should be said about the daring, indeed the *audacity*, in the way Dr. Nordhaus has approached important challenges. I'm going to focus on that.

Let's start with Bill's life-changing (and world-changing) decision to apply economic analysis to the phenomenon of climate change. The decision met with surprise and resistance. From the mid '70s to the late '80s, Bill spent a lot of time at the International Institute of Applied Social Analysis (IIASA), a research institute devoted to a range of global problems. Bill had been offering advice on IIASA's model of the global energy system. It was there that Bill first sought to view the climate system through an economic lens. He took some initial steps to introduce the climate system in IIASA's energy model, which hitherto included no climate component. Not long after Bill began this effort, the leader of the energy model's program, a nuclear physicist, ordered Bill to stop the work, seeing little value to a focus on climate. (Nuclear power issues, on the other hand, were very important.) Undeterred, Bill went to the director of IIASA and got permission to focus on CO_2, largely on his own. What initially was to be a new component of IIASA's energy model became Bill's own baby.

This chapter was originally published in Climate Change Economics, Vol. 11, No. 4, December 2020, published by World Scientific Publishing, Singapore. Reprinted with permission.

Of course, the idea startled economists as well. About 30 years ago, when Bill disclosed at a Brookings conference that he'd begun to do research on the economics of climate change, the reaction was puzzlement and skepticism. This type of reaction was typical early on. Climate change was not on economists' maps, and addressing it with economic tools seemed bizarre. It takes courage to risk ridicules from your colleagues.

It was courageous enough for Bill simply to decide to apply economics to the climate change problem. But the manner in which he chose to represent the climate system was equally bold. Around 1990, Bill approached climate scientist Stephen Schneider at the National Center for Atmospheric Research. NCAR was perhaps the leading US institution involved in the development of climate models: models that connected CO_2 emissions to changes in CO_2 concentrations in the atmosphere and oceans, and linked these changing concentrations to changes in global surface temperature and in patterns of precipitation. These climate models were really, *really* big — often with (literally) millions of equations. Bill indicated to Schneider that he wanted to include the climate system in his new model. Specifically, he mentioned — and here's the audacity — that he wanted his new integrated climate-economy model to capture the climate system . . .*in less than a dozen equations*. Years later, Schneider confessed to me that initially he questioned Nordhaus's sanity. It attests to Schneider's breadth of intellect that he soon came to appreciate Bill's unusual aspirations. He helped Nordhaus come up with the eight-equation climate system in the original DICE model.[1] The two developed a lasting professional connection as well as a strong friendship.[2]

Then there's Bill's taking transparency to the extreme. When introducing his very young DICE model in workshops in the early '90s, he would display the model's equations and state, without apology, "It's a model you can write down on one page." This was in a decade in which enormous advances in computer memory capacity and computation speed had spawned the production of very large, high-dimensioned macroeconomic models. Bill had the courage to buck the macro model trend: the complete DICE model consisted of 14 equations. Was this a virtue? It was, big time. DICE's transparency has helped researchers around the world reproduce the model as well as modify or extend it. This has conferred huge benefits to the research and policy communities. Beyond transparency, Bill's work exemplifies accessibility. In addition to the model's structure, DICE's computer code and all the data have been fully disclosed from the beginning, routinely included in Bill's papers as well as his first DICE book.[3] So, with minimal effort, researchers have been able to replicate the results

---

[1] See Table 2.2 of Nordhaus's *Managing the Global Commons: The Economics of Climate Change*. Cambridge, Mass., MIT Press, 1994.

[2] Bill admires and speaks very fondly of Schneider, who died in 2010 of lymphoma. He credits Schneider and his collaborator Starley Thompson with a key breakthrough that made climate-change modeling possible: the reliance on radiative forcing as the bridge between CO_2 concentrations and temperature change.

[3] Nordhaus, *Managing the Global Commons, op. cit.*

and explore how changes in data, parameters, or structure influence the outcomes. Enormously beneficial to climate policy researchers!

Let us not forget the importance of a name. I can't claim that Bill revolutionized model-naming, but he sure displayed a talent for it. His name for the climate-economy model — "Dynamic Integrated Climate-Economy Model" — is brilliant. The associated acronym sticks to your memory and yields an image that connotes a central feature of the climate change problem: uncertainty. Am I going overboard here, conferring too much importance to a name? I ask you to consider if instead Bill had christened the work, "Model Connecting the Climate System and the Economy." The acronym is unpronounceable. Referring to the model would have been a chore. Showing appreciation where it is well deserved, Bill has remarked that he's always loved the model's name.[4]

Another display of boldness relates to Bill's seminal paper with Rob Mendelsohn and Daigee Shaw on the impact of climate change on agriculture.[5] Breaking with tradition for economics research papers at the time, the paper presented key results in color. It contained multi-colored maps showing the projected impact of climate change on the profitability of agriculture across counties throughout the continental US. Somehow, the authors persuaded the *AER* to retain the color in the published article. I believe this was the first time the *AER* printed one of its articles in color. Isn't this accomplishment as important as the insights offered in many highly cited economics papers?

For my final example of daring, I'll venture outside of the climate economics domain. For economists who wish to measure real economic growth, it is essential to get price indexes right, since these indexes are needed to express the actual value to consumers of the goods and services that can be enjoyed from given nominal wages or incomes. This is not a trivial challenge because production methods, product quality, and product variety change a great deal over time, complicating the translation from goods to value (money-equivalent of utility). Around 1990, when Bill was concentrating hard on the price-index problem, he surmised that existing studies failed to capture all of the dimensions that lead to reductions over time in real prices — and that as a result these studies understated the reductions in real prices and understated economic growth. To address this issue empirically, he focused on illumination services — that is, light. Existing studies typically employed data that spanned a century at most, but Dr. Nordhaus decided to consider a time-span of ... *1.4 million*

---

[4]In an unpublished note, and with considerable flourish, Bill indicates how the acronym connects with the formidable trade-offs that characterize the climate problem: "[The word] DICE also conveys a shiver of risk and danger. It alluded to the Faustian bargain that we make as we continue down the path of unchecked climate change, the Walpurgis Night of reveling without reckoning on how the devil of damages will come to drag us to a hellish future."

[5]Mendelsohn, Rt, WD Nordhaus and D Shaw (1994). The impact of global warming on agriculture: A Ricardian analysis. *American Economic Review* 8(4), 753–771.

years. The well documented time series he developed stretched from the production of light by fire (in caves by our early human ancestors) to lamps, candles, and compact fluorescents. Needless to say, this gigantic departure from traditional empirical work took one helluvalotta nerve. The results from the study[6] are as stunning as the length of the time-series. Bill estimated that the energy efficiency of light production — lumens per watt expended — had increased by a factor of about 30,000 since prehistoric times. After accounting for these enormous efficiency improvements as well as the changing characteristics of light over time, Bill arrived at what he considered to be a truer measure of the price of light. According to his measure, from 1800 to 1992, the real price of light declined by a factor of over 900, while a conventional approach as

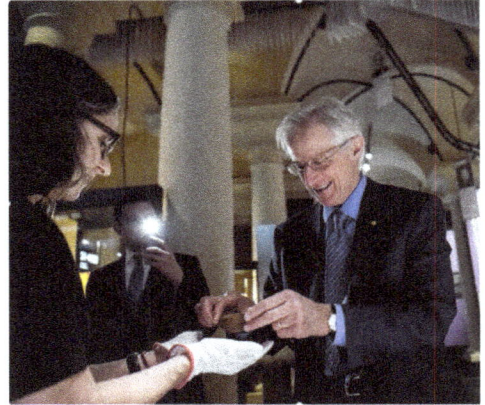

defined in the paper would indicate it declined by a factor of just 3–5. The paper made a compelling case that for other goods and services, existing studies had seriously understated the extent to which their real prices had declined — and thus had understated real wage and real income growth.

Bill is most proud of this paper. Every Nobel laureate is asked to make a gift to the Nobel Foundation that represents his or her contribution to his or her Science. Bill gave them a lamp from antiquity, an allusion to his Light paper. It is shown above.

One wonders whether Bill would have won the Nobel, had he not been extraordinarily daring as well as brilliant. I'm sure Bill wouldn't shy away from answering this question.

Thanks, Bill, for your audacity.

Acknowledgments

Thanks to Lint Barrage, John Weyant, and Gary Yohe for incriminating evidence.

---

[6]Nordhaus, WD (1998). Do real output and real wage measures capture reality? The history of Light suggests not. Cowles Foundation Paper No. 957, Yale University. Reprinted in RJ Gordon and TF Bresnahan (eds.), *The Economics of New Goods*. University of Chicago Press, pp. 27–70.

CHAPTER 4

INTEGRATED ASSESSMENT AND CLIMATE CHANGE

ROBERT MENDELSOHN

Yale School of the Environment, 195 Prospect Street
New Haven, CT 06511, USA
robert.mendelsohn@yale.edu

The crowning achievement of the many published papers and books that William Nordhaus has published on climate change is the development of a simple Integrated Assessment Model of climate change. Embedding natural science insights into an economic framework reveals one can "solve" this difficult problem "for the greatest good, for the greatest number, and for the longest time". Making certain that all the pieces are empirically based, and fit tightly together, and are internally consistent reveals this to be a masterpiece in the fine art of Integrated Assessment.

Keywords: Integrated Assessment Modeling; climate change.

1. Introduction

Although academic research has driven economics and natural science forward, academics are often poorly suited to address complex problems like climate change involving a host of disciplines. Faculty tends to be narrowly organized into disciplines with few incentives to work with other disciplines. Each discipline has an important contribution to make to understand the chain of events that greenhouse gas emissions set in motion, but no single discipline has the breadth of knowledge to understand all the links between actions taken to limit greenhouse gases and the final consequences. Many individual disciplines have critical insights about specific phenomena, but they are often unaware of the insights of other disciplines. When a complex problem such as climate change appears, academics claim it is a "wicked problem" too difficult to solve but it may be simply too wicked for an individual discipline to solve on its own. Integrated Assessment (IA) organizes the vast insights of multiple disciplines to create a unified and internally consistent composite vision. It is a shining example of the whole being far greater than the parts from which it is made. The single most important contribution that Prof. Nordhaus made to climate change is to build the first Integrated Assessment Model (IAM) of climate change that embedded the essence of the natural

This chapter was originally published in Climate Change Economics, Vol. 11, No. 4, December 2020, published by World Scientific Publishing, Singapore. Reprinted with permission.

sciences into a macroeconomic framework (Nordhaus, 1991). He showed that the wicked problem could be solved.

In Section 2, we discuss the importance of certain natural science and economic insights that are essential to understanding climate change. This is one of the key strengths of IAMs. One does not necessarily need a highly detailed IAM to study climate change. The DICE model has surprisingly little detail. But there are some critical features in the model that cannot be left out. Although alternative assumptions about climate change can change the mitigation results of a model, if studies do not capture the key insights of the underlying science, the study likely provides little insight into managing greenhouse gases.

Section 3 of the paper discusses the importance of internal consistency. The beauty of Integrated Assessments when they are done properly is that they are internally consistent depictions of outcomes. Without the framework of Integrated Assessment forcing this internal consistency, groups of disciplines often make different assumptions that are inconsistent with the assumptions of other disciplines. Although these assumptions can be readily altered to be consistent, it is difficult to coordinate across multiple disciplines so that they are all playing by the same rules. Assumptions can conflict. Without a clear structure, multiple disciplines can cause analyses to be inconsistent with each other and sometimes obscure a clear picture of likely or preferred outcomes. For example, the IPCC process has led climate science, impact sciences, and mitigation cost assessments to be made independently. The result in AR5 was that the climate and impact assessment assumed Business as Usual (BAU) was the RCP8.5 scenario (IPCC, 2013, 2014a) which requires a GDP per capita growth rate of 3% per year whereas the mitigation cost assessment (IPCC, 2014b) assumed Business as Usual was equivalent to an emission scenario half way between RCP6 and RCP8.5 implying a future GDP per capita growth rate of 2% per year. The mitigation cost assessment was consequently not comparable with the climate and damage assessment (Mendelsohn, 2016).

Section 4 of this paper argues that IAMs have underestimated the importance of adaptation. Admittedly, it is easier to simply estimate potential damage assuming that there is no human reaction to climate damages and benefits. However, humans have adapted to climate since they first existed so it is unrealistic to assume that they will suddenly stop. If they efficiently adapt, they will trim away some of the potential climate damage that many impact models predict and they will seize some of the benefits that climate impact models ignore. The actual damages will be lower than potential damage. In some cases, such as sea level rise, adaptation will lower potential damage by an order of magnitude (Diaz, 2016).

The concluding section of this paper (Sec. 5) tries to place IAMs in perspective as a policy tool. Integrated Assessments provide insight into how global choices can be optimized. But do IAMs capture every element of climate change? What is missing? If they are biased, in what direction is the bias? Is it useful to provide insights concerning how climate change can be managed in a perfect world when the real world is anything but perfect? No IAM is perfect, but partly IAMs will always be controversial because they reflect a global choice, an aggregate of global values. Critics of IAMs will always rail against IAM results that are inconsistent with their own personal preferences. But does this controversy suggest that IAMs have a limited effectiveness or is the controversy in fact a tribute to the importance of IAMs?

2. Integrating Natural Science into Economic Models

One of the principle purposes of climate change IAMs is to capture how greenhouse gas emissions affect society over time. This problem involves a host of natural science disciplines, each of which contributes a link between emissions and consequences. The problem is daunting because there are many links, the entire earth is at play, and the time horizon between actions and consequences plays out over millenniums. What must be included in a model of this process and what is not that important? The following discussion follows the general logic of the AR5 report on climate change (IPCC, 2013, 2014a, 2014b).

The first link in the science model is between greenhouse gas emissions and global concentrations. One might assume that the stock of greenhouse gases in the atmosphere is simply the sum of historic emissions. However, models of global geochemical cycles of individual greenhouse gases make clear that the atmospheric stock is a product of not just cumulative manmade emissions but also natural emissions and sinks (IPCC, 2013). As the stock of each gas in the atmosphere increases, it changes the rate of natural emissions and sinks. For example, increases in the atmospheric stock of carbon dioxide (CO_2) increases their absorption into the ocean. Higher CO_2 levels cause plants and especially trees to grow faster causing the absorption of CO_2 from the atmosphere into land to increase. How quickly CO_2 and other greenhouse gases move from the atmosphere back into these other stocks describes the dynamics of these natural systems and how long the gases will remain in the atmosphere naturally. Thawing tundra may release large quantities of methane. These feedbacks from the global biochemical cycles are critical to link how emissions alter concentrations.

The next critical link is between concentrations and radiative forcing (IPCC, 2013). Each gas has a different ability to block infrared radiation. The combination of how long a molecule stays in the atmosphere and how much it increases radiative forcing explains the relative greenhouse gas effect of that specific gas. Radiative forcing gradually heats up the ocean. It is the warming of the ocean that leads to climate change. The key role the ocean plays explains why climate models are Atmospheric Oceanic Global Circulation Models. The warming of the deeper parts of the ocean is the link between cumulative radiative forcing and climate change. It takes centuries (possibly millenniums) for a given level of greenhouse gas concentrations to reach an equilibrium temperature with the oceans. Early climate research in the 1990s focused on measuring equilibrium temperature effects, suggesting a doubling of greenhouse gas concentrations would increase global temperatures by 3°C. However, the equilibrium temperature that will be reached in many centuries has only a small effect on the present value of near-term mitigation benefits. The critical question is how temperatures will change over the next 100 years. Climate models have become far more adept at measuring these transient temperatures. Doubling greenhouse gases will only increase temperatures by 2100 by 2°C above today's temperature. Properly accounting for the timing of transient temperature change is critical to balancing the costs and benefits of mitigation.

The melting of the vast sea ice near the North Pole has exposed the northern atmosphere to warm oceans underneath. This explains why the regions of the planet near the North Pole are warming so much faster than the rest of the planet (IPCC, 2013).

Another important consequence of global warming is sea level rise (IPCC, 2013). The primary source of current sea level rise is from the expansion of ocean water as it warms. There is also a small contribution from land-based glaciers and ice sheets. However, there remain vast amounts of water in the ice sheets in Greenland and Antarctica (25 m of ocean) (IPCC, 2013). What is uncertain about the ice sheets is how quickly they will melt. Will they melt away by 2500 or will they take many more centuries? This is another critical uncertain component of the science. A slow melt implies a manageable gradual increase in sea level rise but a rapid melt implies the potential destruction of vast capital along global coastlines.

As the planet warms, there will be other changes that natural science models predict. The increase in temperature will speed up the hydrological cycle leading to more water in the atmosphere which will cause the planet to warm even more (a positive feedback that the climate models incorporate) (IPCC, 2013). Evaporation rates and rainfall will increase. How precipitation patterns will change across the Earth's surface, however, is uncertain. It will depend on how climate change alters wind patterns. Although the aggregate amount of rainfall is expected to increase, global climate models do not agree on how precipitation will be distributed. Some places may get more rainfall while other places will get less. The distribution of rainfall across seasons may well also change.

In addition to changes in temperature and precipitation, extreme events such as tropical cyclones are expected to become more powerful (Emanuel *et al.*, 2008). These massive storms in turn would lead to substantial coastal and inland flooding as well as damaging high winds. Other storms may also change their behavior.

Ecosystems are predicted to change as temperatures change (IPCC, 2014a). The effect of warming and higher GHG concentrations over the next 150 years is expected to cause natural forests to expand toward the poles replacing tundra and ice (Sitch *et al.*, 2008). Tropical forests will move into areas currently occupied by warm temperate forests. Temperate forests will gradually replace boreal forests and what is left of boreal forests will move into tundra. If temperatures exceed 5°C, the expansion of global forests will end and many forests will gradually turn to parkland and savannah. Overall, deserts will expand in a warmer world in places where there is insufficient rainfall (Sitch *et al.*, 2008). Animals are expected to track habitat as climate changes, provided they are not blocked by natural or manmade barriers. Animals suited for tropical and temperate forests and savannah and parkland should be able to endure climate change as their habitat expands. Of course, some living things may have more difficulty moving with their habitat than others and may not be able to make the transition on their own. Animals and other living things that live in ecosystems that are contracting such as tundra and boreal forest, will likely shrink in numbers as their habitat shrinks (IPCC, 2014a).

A small fraction of the world's economy is directly sensitive to climate (IPCC, 2014a). Agriculture, forestry, water-related industries, coastal cities, energy demand and supply, tourism, and some infrastructure are climate-sensitive. If the world did not spend more on cooling, there is evidence that labor productivity could fall with warming (Adhvaryu *et al.*, 2020; Deschenes and Greenstone, 2011). As climate changes, climate-sensitive sectors either thrive or more often are damaged. The size of these future damages will depend on the future growth rates of these climate-sensitive sectors. Because some of these climate-sensitive sectors such as agriculture are a declining fraction of future GDP, impacts to these market sectors may well shrink as a fraction of GDP. On the other hand, expenditures on cooling are likely to increase over time because cooling is income-elastic. Cooling damage is likely to increase with higher GDP per capita.

Nonmarket changes from ecosystem changes, changes in disease incidence and health, and extreme events, and climate as an amenity are also important to measure. A great deal is known about some nonmarket changes. Many people place a premium on a mild Mediterranean climate (Maddison *et al.*, 2013). People who live in cold environments often seek much warmer climates for vacation (Hall and Higham, 2013). People also place a significant value on the loss of keystone species such as polar bears and tigers (Ando and Langpap, 2018). But comparatively, little is known about the nonmarket value of ecosystems in general. If boreal forest is replaced by temperate forest or temperate forest by tropical forest, is this a gain or a loss to future residents? As warming allows some disease vectors to move to new places, there will be a cost imposed on future residents from either additional disease or more likely public health measures. A warmer world will entail more frequent exposure to hot temperatures but less frequent exposure to cold temperatures. The net effect on health depends upon which effect is greater. Currently, cold exposure kills about 20 times more people than heat exposure (Gasparrini *et al.*, 2015). Recreation in winter such as skiing will be damaged by warming. On the other hand, summer recreation will expand. Studies in the United States suggest summer recreation far outweighs winter recreation (Mendelsohn and Markowski, 1999). A complete accounting of nonmarket effects is another important feature to include in IAMs.

Including a nonlinear damage function of these impacts was another important innovation by Nordhaus in order to capture key long-term dynamic aspects of climate change (Nordhaus, 1992). With a nonlinear damage function, the problem gets worse as temperatures rise. That leads to a dynamic outcome where the shadow price of carbon rises with the stock of greenhouse gases in the atmosphere. As the stock accumulates and the marginal damage per ton of emission increases, the carbon price rises over time. This insight led to the extensive literature on the social cost of carbon (e.g., Nordhaus, 2014, 2017).

Nordhaus has gone on to write and edit several books developing Integrated Assessment Models of climate change (Nordhaus, 1994, 2008, 2013; Nakicenovic *et al.*, 1994, 1996; Nordhaus and Boyer, 2000). He developed many of the key insights that economics has offered to the climate change debate. For example, Prof. Nordhaus along with Yang developed a regional model of climate change (RICE) that began to explore the heterogeneity of the world (Nordhaus and Yang, 1996). As a classic public good, the heterogeneity of impacts implies countries will find it difficult to agree on a single global temperature target. Polar countries will want a warmer target and low-latitude countries will want a cooler target. There is also a heterogeneity of mitigation costs across countries as nations with substantial fossil fuels may be reluctant to leave them in the ground. Finally, there is the selfish incentive that every country is better off letting others pay for mitigation. These combined problems have not been resolved across nations to the detriment of any real progress on global mitigation.

Many authors have now contributed to a host of climate IAMs, each designed to address different aspects of the climate problem. Some of these models, like DICE and RICE, maximize the present value of welfare but have added more detail about regional differences, technologies, and impacts (Manne *et al.*, 1995; Bosetti *et al.*, 2009; Anthoff and Tol, 2010). Other IAMs start with cumulative emission targets as a goal and solve backward to find a least cost solution (Jacoby *et al.*, 2006; Calvin *et al.*, 2019; Fujimori *et al.*, 2017; Huppmann *et al.*, 2019). The cumulative emission target becomes a nonrenewable resource and the carbon price rises with the interest rate over time. The focus of these models is not on what target is best but rather on the least cost package of technologies over time that lead to the desired long-term target. By equating the marginal cost of each technology to the carbon price in each period, these models determine what mitigation strategies need to be in place each decade. Some of these models explore whether all these technologies could fit on the planet at the same time given the required energy demand, land, and water available over time (the energy, land, and water nexus).

One of the basic insights of all these models is that the cost of reaching ever more stringent targets rises rapidly as the target tightens. For example, a target of 2.5°C in 2100 is expected to cost between $6 trillion/yr and $22 trillion/yr in 2100 but a target of 1.6°C is expected to cost between $15 trillion/yr and $57 trillion/yr by 2100 (IPCC, 2014b). The IAMs also make clear that one has to use all the available cost-effective measures to control carbon. It gets increasingly more expensive not to use nuclear, biomass energy, carbon capture and storage, or forest sequestration. Finally, all the IAMs reveal that delaying the start of mitigation another decade makes stricter targets ever more expensive to reach, eventually leaving only higher temperature targets still on the table. For example, the delay in global mitigation from 1990 through 2020 has made reaching the 2°C target incredibly expensive.

3. Internal Consistency

One of the most important potential advantages of an Integrated Assessment Model is that it can be internally consistent. All the assumptions within the model can be designed to be consistent with one another. One beautiful feature of DICE is its internal consistency. However, as models get more complicated and feedbacks grow, it can be very difficult to keep a model internally inconsistent. There are simply too many moving parts that have to be designed by people unfamiliar with the rest of the model. Many IAM models may still lack internal consistency. If model builders are content with the results, the details are often not reviewed. The inner workings of most models are hidden and papers using IAMs commonly treat them as a black box. The lack of external review of many models means that hidden errors are less likely to ever be identified.

Sometimes assumptions about natural constraints are inconsistent, for example, so that models predict outcomes which exceed the available regional land or water resources. But models of natural science have gotten ever more sophisticated over time so that regional limits are now respected in at least some Integrated Assessments. The most disconcerting inconsistencies (perhaps just to an economist) that remain in current IAM studies of climate change concern inconsistent assumptions about the economy.

One of the least understood benefits of DICE is that it incorporates a macroeconomic model into the analysis of climate change. The macroeconomic model reveals that population growth rates and GDP per capita growth rates are the drivers of the path of interest rates and BAU greenhouse gas emissions. Large emission scenarios require fast economic growth rates. It makes no sense, as many studies have done, to combine arbitrary socioeconomic scenarios with rapid emission scenarios such as RCP8.5. Low and medium growth rates simply cannot produce that many emissions. Specific growth rates of population and GDP per capita lead to specific interest rates and BAU emissions over time. Examining scenarios that are internally inconsistent does not shed light on future possibilities because these scenarios are not possible. Mixing them with scenarios that are likely only clouds our understanding of the problem and its solution.

If GDP per capita grows rapidly over the next century, the interest rate will be high. The current generation will be the poorest in the century and will not want to sacrifice deeply for much richer future generations. The current generation will need a large return to forego current consumption. The high rate of GDP growth, in turn, will also lead to a strong growth in energy demand. If fossil fuels remain the cheapest source of energy, this will lead to a rapidly growing BAU emissions path. This in turn leads to rapidly growing temperatures. If the economy is very sensitive to higher temperatures, this will curb far-future growth and therefore curb far-future potential emissions.

The scenario also implies that future carbon prices will be high stimulating more mitigation and even lower actual emissions.

In contrast to the above scenario, if GDP per capita grows slowly, the interest rates will be lower. Future generations will not be much richer than the current one. The low GDP growth rate implies little increase in energy demand and therefore slowly growing BAU emissions. Temperature will slowly increase as will future incomes, so climate impacts will be relatively small. Carbon prices will be lower and there will be less mitigation.

When climate change analyses mix and match interest rates, growth rates, and emission paths, they are effectively combining scenarios with zero probability of occurrence with scenarios that are more likely than they recognize. Without properly weighing the probabilities of each scenario, they are adding noise and often biasing the outcomes.

4. Adaptation

There is a great deal of confusion about adaptation in the climate literature which has led to many IAMs underestimating adaptation. Part of the confusion concerning adaptation is that many analysts assume adaptation is very similar to mitigation. Adaptation, however, does not require international cooperation, it does not require the same long-term perspective as mitigation, and its sole purpose is to make the person(s) who does (do) the adaptation better off. People and firms already have a selfish incentive to do adaptation whereas their selfish tendencies would keep them from doing mitigation. The purpose of adaptation is to make each individual party better off given that their local climate is changing. Adaptation should only be undertaken if it leads to net benefits to the target population. If the cost of an adaptation exceeds the benefit, it is a maladaptation and it should not be done. Often times, the benefit is going to be a reduction in climate damage but sometimes it is seizing a new opportunity that climate change creates. For example, farmers should change crops as climate changes if the change leads to higher net revenue per ha. Households should invest in air conditioning if the resulting comfort is worth more to them than the cooling costs. Private adaptation does not need to be subsidized to happen nor does private adaptation need government regulations.

Public adaptation benefits a collection of people. Collections of people often find it difficult to coordinate and pay for things that benefit them all. Public adaptation does need government assistance (Mendelsohn, 2006). But a lot of pubic adaptation can be executed by existing local governments. The local government already has an incentive to engage in activities that benefit local citizens. Large river systems can be managed by regional governments. Some public health efforts such as controlling pandemics can be handled if effective national governments actively assist. Many governments have proven to be inept at providing important public goods, but the institution exists. Governments may need technical help learning how to adapt, especially to something they are not yet used to managing. But public adaptations can largely be done by each government helping its own constituents.

The purpose of adaptation is to maximize net benefits, not to reduce climate damage at all costs. The reduced climate damage must be greater than the increase in costs or the program will just make the society even worse off. Individuals, firms, and governments already have all the incentives they need to choose desirable adaptations.

Adaptations do not need to last indefinitely. Some adaptations such as adding more water to a crop might last just a growing season. Adaptations that involve capital investments need only last for the lifetime of the investment. Short-lived investments intended to last only five or ten years, need only to be effective over this limited time period. Simply knowing the current climate is sufficient. Long-lived investments that might last several decades, do have to be concerned about the future climate over coming decades but usually a good forecast of immediate climate changes is sufficient. Adaptation is not like mitigation where the benefits last centuries. Adaptations can be done with just limited foresight of the climate in the near term.

All of these insights into adaptation suggest that it is very likely that adaptation is going to be done. Adaptation will lower the net actual damage caused by climate change below the potential damage. It is important for IAMs to include adaptation because they will otherwise overestimate climate damage and call for too much mitigation. In particular, adaptation makes a huge difference to the magnitude of damage in agriculture, sea level rise, water, and public health. Changes by farmers in crop choice, livestock choice, timing, and irrigation can substantially reduce agricultural damage (Mendelsohn and Dinar, 2009). Defending urban areas but retreating in rural areas of the global coastline will reduce the damage from sea level rise by an order of magnitude (Diaz, 2016). Reallocating water from low-valued irrigation to high-valued residential and industrial uses will reduce water damage (Hurd *et al.*, 1999, 2004). Providing inexpensive public health measures to curb vector-borne diseases is a lot cheaper than letting poor people die (Pandey, 2010). The increased expenditures for cooling already counted as a damage in the energy sector will sharply reduce potential deaths and lost productivity from heat waves (Deschenes and Greenstone, 2011).

5. Conclusion

This paper argues that there are two important elements that make Integrated Assessment Models a valuable tool for studying climate change. First, it integrates the insights of a multitude of disciplines into a single unified tool. The combined insights of all these disciplines are far more powerful and wise than any model from a single discipline can be for a complex problem such as climate change. Second, it encourages the analyses across all these disciplines to be consistent so that together they reproduce the logical chain of events linking actions such as emissions to consequences.

IAMs often take an ideal world and assume an ideal reaction to climate change. For example, on the cost side, the mitigation cost estimates assume that the entire world will equate the marginal cost of mitigation across all polluters efficiently over time. That implies a coordinated global carbon price that rises with damage over time and universal cooperation amongst all governments and across all emitters. There is no mitigation plan yet in existence that lives up to this ideal. It is still useful to see what one could accomplish if such a system were created. If actual systems are going to be much less effective, IAMs can still provide useful feedback by showing how much more expensive mitigation will be if it is done less than perfectly. More costly mitigation systems will not be able to control emissions as much as a perfect system. Temperature will rise higher. But just because the world is not perfect does not imply one is no longer interested in modeling and planning.

Of course, models by design are simplifications of more complex underlying phenomena. Models will always leave out details. Simple models may lead critics to complain that models have not quantified the important impacts. How the models have handled nonmarket impacts is particularly problematic because many of these impacts are poorly understood and not effectively valued. Whether these criticisms are valid depends on whether a solid case can be made that careful measurements would change the results.

Of course, each part of an IAM model can potentially fail to capture the critical nature of what it is reproducing. A basic tenet of modeling is that the assumptions of the model need empirical justification. Modelers are not free to make arbitrary assumptions. Model forecasts should reflect our best understanding of the world. Further, no matter how carefully a model is constructed, model forecasts will always be uncertain. Models are not exact reproductions of the future, they simply reflect our current understanding. But the models help focus what we know about climate change and what we do not. Critics that feel a model has not captured an important element can argue their case. For example, this paper argues that most IAMs have underestimated adaptation. Ongoing debate about model elements will continue to be useful just as competitive forces are helpful in keeping markets on track.

Finally, some critics argue that IAM models should be disregarded because they are controversial. However, this is a logical error. IAM models are controversial because they are important and each model tends to point policy in a specific direction. The policies are controversial because they make global recommendations about desirable paths of mitigation and therefore a desired global outcome. However, there is no single desired global temperature that is best for everyone. People in low latitudes which are relatively hot would prefer a cooler planet. People in the temperate zone would prefer the planet stay as it is. People in cold polar regions would benefit from a warmer planet. Coming to a common agreement on a global temperature and a global method to reach it does not imply each person gets exactly what they individually want. A common agreement will never please everyone. No matter what choice is made, it will be in some people's personal interest to pick a different choice. There will always be controversy about global choices, so any model which leads to a specific choice should be expected to endure criticism.

William Nordhaus has demonstrated that a relatively simple IAM can capture the essential elements of a climate change and be solved. It remains to future generations to uncover what was not first understood about climate change as the world evolves and to modify the model to include missing pieces. The solution may well change as our understanding grows. This is a fundamental skill of humans that has helped them adapt in the past. There is every reason to believe future generations will continue to adapt with skill and cunning.

Acknowledgments

I wish to thank David Maddison for his insightful and balanced comments on an earlier draft.

References

Adhvaryu, A, N Kala and A Nyshadham (2020). The light and the heat: Productivity co-benefits of energy-saving technology. *The Review of Economics and Statistics*. doi: 10.1162/rest_a_00886.

Ando, A and C Langpap (2018). The economics of species conservation. *Annual Review of Resource Economics*, 10, 445–467.

Anthoff, D and R Tol (2010). On international equity weights and national decision making on climate change. *Journal of Environmental Economics and Management*, 60(1), 14–20.

Bosetti, V, C Carraro, E Massetti, A Sgobbi and M Tavoni (2009). Optimal energy investment and R&D strategies to stabilize greenhouse gas atmospheric concentrations. *Resource and Energy Economics*, 31(2), 123–137.

Calvin, K, P Patel, L Clarke, G Asrar, B Bond-Lamberty, RY Cui, A Di Vittorio, K Dorheim, J Edmonds, C Hartin, M Hejazi, R Horowitz, G Iyer, P Kyle, S Kim, R Link, H McJeon, SJ Smith, A Snyder, S Waldhoff and M Wise (2019). GCAM v5.1: Representing the linkages between energy, water, land, climate, and economic systems. *Geoscientific Model Development*, 12(2), 677–698.

Deschenes, O and M Greenstone (2011). Climate change, mortality, and adaptation: evidence from annual fluctuations in weather in the US. *American Economic Journal: Applied Economics*, 3(4), 152–185.

Diaz, DB (2016). Estimating global damages from sea level rise with the Coastal Impact and Adaptation Model (CIAM). *Climatic Change*, 137, 143–156.

Emanuel, K, R Sundararajan and J William (2008). Tropical cyclones and global warming: Results from downscaling IPCC AR4 simulations. *Bulletin American Meteorological Society*, 89, 347–367.

Fujimori, S, T Masui and Y Matsuoka (2017). AIM/CGE V2.0 model formula. In *Post-2020 Climate Action*, S Fujimori, M Kainuma and T Masui (eds.), pp. 201–303. Singapore: Springer.

Gasparrini, A, Y Guo, M Hashizume, E Lavigne, A Zanobetti, J Schwartz, A Tobias, S Tong, J Rocklov, B Forsberg, M Leone, M De Sario, ML Bell, Y-LL Guo, C-F Wu, H Kan, S-M Yi, M de Sousa Zanotti Stagliorio Coelho, PHN Saldiva, Y Honda, H Kim and B Armstrong (2015). Mortality risk attributable to high and low ambient temperature: A multicountry observational study. *The Lancet*, 386(9991), 369–375.

Hall, CM and J Higham (eds.) (2003). *Tourism, Recreation, and Climate Change.* Clevedon, UK: Channel View Publications.

Huppmann, D, M Gidden, O Fricko, P Kolp, C Orthofer, M Pimmer, N Kushin, A Vinca, A Mastrucci, K Riahi and V Krey (2019). The MESSAGE Integrated Assessment Model and the *ix modeling platform* (ixmp): An open framework for integrated and cross-cutting analysis of energy, climate, the environment, and sustainable development. *Environmental Modelling & Software*, 112, 143–156.

Hurd, B, M Callaway, JB Smith and P Kirshen (1999). Economic effects of climate change on US water resources. In *The Impact of Climate Change on the United States Economy*, R Mendelsohn and J Neumann (eds.), pp. 133–177. Cambridge, UK: Cambridge University Press.

Hurd, BH, M Callaway, J Smith and P Kirshen (2004). Climatic change and US water resources: From modeled watershed impacts to national estimates. *Journal American Water Resource Association*, 40(1), 129–148.

IPCC (2013). *AR5 Climate Change 2013: The Physical Science Basis — Contribution of Working Group I to the Fifth Assessment Report of the Intergovernmental Panel on Climate Change*, TF Stocker *et al.* (eds.). Cambridge, UK and New York, NY: Cambridge University Press.

IPCC (2014a). *AR5 Climate Change 2014: Impacts, Adaptation, and Vulnerability — Contribution of Working Group II to the Fifth Assessment Report of the Intergovernmental Panel on Climate Change*, CB Field *et al.* (eds.). Cambridge, UK and New York, NY: Cambridge University Press.

IPCC (2014b). *AR5 Climate Change 2014: Mitigation of Climate Change — Contribution of Working Group III to the Fifth Assessment Report of the Intergovernmental Panel on Climate Change*, O Edenhofer *et al.* (eds.). Cambridge, UK and New York, NY: Cambridge University Press.

Jacoby, HD, JM Reilly, JR McFarland and S Paltsev (2006). Technology and technical change in the MIT EPPA model. *Energy Economics*, 28(5–6), 610–631.

Maddison, D, K Rehdanz and D Narita (2013). The household production function approach to valuing climate: the case of Japan. Climatic Change, 116(2), 207–229.

Manne, A, R Mendelsohn and R Richels (1995). MERGE: A model for evaluating regional and global effects of GHG reduction policies. *Energy Policy*, 23, 17–34.

Mendelsohn, R and M Markowski (1999). The impact of climate change on outdoor recreation. In *The Impact of Climate Change on the United States Economy*, R Mendelsohn and J Neumann (eds.), pp. 267–288. Cambridge, UK: Cambridge University Press.

Mendelsohn, R (2006). The role of markets and governments in helping society adapt to a changing climate. *Climatic Change*, 78, 203–215.

Mendelsohn, R and A. Dinar. 2009. *Climate Change and Agriculture: An Economic Analysis of Global Impacts, Adaptation, and Distributional Effects*, Edward Elgar Publishing, Cheltenham, UK.

Mendelsohn, R (2016). Should the IPCC assessment reports be an integrated assessment? *Climate Change Economics*, 7(1), 1640002:1–1640002:8.

Nakicenovic, N, WD Nordhaus, R Richels and FL Toth (eds.) (1994). Integrative assessment of mitigation, impacts, and adaptation to climate change. Report No. CP-94-009, International Institute for Applied Systems Analysis (IIASA), Vienna.

Nakicenovic, N, WD Nordhaus, R Richels and FL Toth (eds.) (1996). Climate change: Integrating science, economics, and policy. Report No. CP-96-1, International Institute for Applied Systems Analysis (IIASA), Vienna.

Nordhaus, WD (1991). To slow or not to slow: The economics of the greenhouse effect. *The Economic Journal*, 101, 920–937.

Nordhaus, WD (1992). An optimal transition path for controlling greenhouse gases. *Science*, 258(5086), 1315–1319.

Nordhaus, WD (1994). *Managing the Global Commons: The Economics of Climate Change.* Cambridge, MA: The MIT Press.

Nordhaus, WD (2008). *A Question of Balance: Weighing the Options on Global Warming Policies.* New Haven, CT: Yale University Press.

Nordhaus, WD (2013). *The Climate Casino: Risk, Uncertainty, and Economics for a Warming World.* New Haven, CT: Yale University Press.

Nordhaus, WD (2014). Estimates of the social cost of carbon: Concepts and results from the DICE-2013R model and alternative approaches. *Journal of the Association of Environmental and Resource Economics*, 1(1–2), 273–312.

Nordhaus, WD (2017). Revisiting the social cost of carbon: Estimates from the DICE-2016R model. *Proceedings of the National Academy of Sciences of the United States of America*, 114(7), 1518–1523.

Nordhaus, W and Z Yang (1996). A regional dynamic general-equilibrium model of alternative climate-change strategies. *The American Economic Review*, 86(4), 741–765.

Nordhaus, W and J Boyer (2000). *Warming the World: Economic Modeling of Global Warming.* Cambridge, MA: The MIT Press.

Pandey, K (2010). Costs of adapting to climate change for human health in developing countries. Discussion Paper 11, World Bank, Washington, DC.

Sitch, S, C Huntingford, N Gedney, PE Levy, M Lomas, SL Piao, R Betts, P Ciais, P Cox, P Friedlingstein, CD Jones, IC Prentice and FI Woodward (2008). Evaluation of the terrestrial carbon cycle, future plant geography and climate-carbon cycle feedbacks using five Dynamic Global Vegetation Models (DGVMs). *Global Change Biology*, 14(9), 2015–2039.

https://doi.org/10.1142/9789811247699_005

CHAPTER 5

DIKES VERSUS WINDMILLS: CLIMATE TREATIES
AND ADAPTATION

SCOTT BARRETT

*Lenfest-Earth Institute Professor of Natural Resource
Economics, School of International and
Public Affairs & Earth Institute, Columbia University
sb3116@columbia.edu*

This paper begins with a tribute to William Nordhaus, focusing on the two questions that have motivated his life's work. The first is by how much carbon dioxide emissions should be reduced over time. The second is how to reach and enforce an agreement among sovereign nations to limit carbon dioxide emissions. Nordhaus was awarded the Nobel Prize for his efforts to answer the first question. I argue here that the answer to this question has been solved to a satisfactory extent, not only by economists, but by diplomats, and that the greatest need now is to answer the second question. I also present a simple model that extends previous research into this second question, a model in which countries choose both whether to abate and whether to adapt. Like all previous research on this topic, including Nordhaus's own, the model doesn't provide a neat solution, only another perspective on one of the most vexing questions in all of human history: how to prevent a tragedy of the commons of global proportions and with profound and possibly catastrophic consequences.

Keywords: International cooperation; free riding; climate change treaty; adaptation.

1. Introductory Remarks on the Contributions of William Nordhaus

William Nordhaus's catalog of contributions to the economics of climate change begins with a working paper completed in 1975, and remains a work in progress.[1] His project, for which he was awarded the Sveriges Riksbank Prize in Economic Sciences in memory of Alfred Nobel, is simple: development of a model that integrates the global human system with the global climate system. In essence, the human system affects the climate; and the climate system's response affects the outcomes attainable by humanity. Unrestrained human behavior, he shows, changes the climate in a way that harms the human enterprise. For its own good, humanity needs to change its ways; most importantly, it needs to reduce its greenhouse gas emissions.

This chapter was originally published in Climate Change Economics, Vol. 11, No. 4, December 2020, published by World Scientific Publishing, Singapore. Reprinted with permission.
[1]This working paper was finally published as Nordhaus (2019b). For a deeper perspective on Nordhaus's contributions, see Barrage (2019).

Nordhaus's preoccupation all these years has been with a single question: By how much should humanity restrain its emissions over time? This is a normative question that can be answered only by incorporating human values. The most important value to Nordhaus's project emerges as a solution of his model: the price associated with the answer to the question of by how much emissions should be limited. In his 1975 paper, Nordhaus (1975, p. 38) calls this value a "shadow price," calculated as "dollars per ton of carbon dioxide emitted into the troposphere." Today we call it the "social cost of carbon."[2]

Though Nordhaus's project is simple, when you contemplate all of the things that must be considered when calculating the social cost of carbon, it is hard not to feel awe at Nordhaus's ambition. To calculate the social cost of carbon you have to know: how much of the CO_2 that is added to the atmosphere will stay there, after accounting for exchanges with the terrestrial biomass and soils, the oceans, and even the Earth's crust, processes that last from days to years, centuries to millennia; how the resulting increase in atmospheric CO_2 can be expected to change global mean temperature over time; how these changes are likely to be distributed spatially and alter precipitation patterns, and how *these* changes in turn will affect ecosystems and critical geophysical systems like ice sheets and ocean currents, all of which may be prone to regime shifts; how all of these changes (translated into measures like wet bulb temperature, sea level rise, and hurricane intensity) affect humans, all over the planet, over long periods of time, net of their efforts to adapt (to include migration and the construction of dikes), which must also be estimated; how the people affected, including future generations living in a future world that is unlike our own, value these changes; and how human population, economic activity, and technology will change over time, partly in response to and partly in anticipation of climate change and the efforts undertaken to limit and adapt to climate change. You also have to recognize that the uncertainties inherent in this exercise are enormous. Climate scientists emphasize uncertainty in the effect a given increase in atmospheric CO_2 will have on global mean temperature (a concept that Nordhaus incorporates in his 1975 paper, long before it had been carefully studied let alone given its current name, "climate sensitivity"), but this is only one of many uncertainties that stand behind estimates of the social cost of carbon. *All* of the relations and values noted above are uncertain, as they must all be calculated for a world that has never existed before. On top of this, you also have to make a judgement as to how humanity values reductions in uncertainty (which is different than humanity's attitude towards risk bearing), brought about by reductions in emissions. Finally, as the social cost of carbon is a normative concept, you have to embed the choice of how much to reduce emissions over time within an ethical framework. You have to decide how to weight the different societies that will be affected by this choice, not only the rich versus the poor and the big versus the small cumulative emitters, but also the big

---

[2]In follow-up work, Nordhaus suggested implementing his solution by universal imposition of "carbon taxes" (Nordhaus, 1977a,b).

versus the small losers from climate change plus any winners. You have to decide whether there should be financial transfers among societies, enabling cost-effective abatement as part of a globally efficient solution. You also have to decide how to weight the "losses" borne by generations that reduce their emissions relative to the "gains" experienced by later generations that, because of these earlier actions, are spared some increment of climate change. Indeed, you even have to consider whether this is the right way to look at the ethics of the choice to reduce emissions, or whether a different perspective should be considered (for example, you might think that the current generation has a moral obligation to the future). To calculate the social cost of carbon, you have to do all of these things, and more. Calculating the social cost of carbon is an audacious undertaking — which explains why Nordhaus's first paper was just a curtain lifter to a lifetime project.

Nordhaus's next contributions on climate change — a second working paper written in 1977 (Nordhaus, 1977a) and a related but shorter paper published in the same year (Nordhaus, 1977b) — take the challenge laid down in the first paper a step further, but stand out to me mainly because they both end by asking a second question: "How can we reasonably hope to negotiate an international control strategy among the several nations with widely divergent interests?"[3] As I said, these papers *end* by asking this question. Neither paper addresses the question. In fact, it wasn't until much later that Nordhaus began to work on it. But he clearly saw back then that the work for which he is most known today, estimation of the social cost of carbon, only had a chance of being implemented if this second question could be answered.

With Zili Yang, Nordhaus made a first attempt to put this second question in context by disaggregating his global model to 10 regions (Nordhaus and Yang, 1996), and calculating the noncooperative and full cooperative time paths for CO_2 emissions and concentrations. This full cooperative time path essentially represents the optimal path that had been the focus of Nordhaus's previous modeling efforts. The novelty in this new paper is estimation of the noncooperative path, a prediction for how countries might behave in the absence of international cooperation. By their estimates, the gap between these two paths — a measure of free riding behavior — is huge, with controls (measured as a carbon tax) being 25 times smaller in the noncooperative outcome than in the full cooperative outcome.[4]

The full cooperative time path for emissions is an ideal, whereas the noncooperative time path for emissions is a kind of default. In his 1975 paper, Nordhaus calculated the "uncontrolled case." In his 1977 paper, he called this the "do nothing" control strategy. So long as humans are organized into nation states, and so long as states possess both the will and the ability to alter the behavior of their citizens for their collective good, they can be expected to "do something." For example, they might adopt the policy advocated by Nordhaus and impose a carbon tax, as a number of countries have done,

---

[3]Nordhaus (1977a, p. 70), Nordhaus (1977b, p. 346).
[4]For more recent estimates of the social cost of carbon for selected large countries, see Nordhaus (2017).

beginning with Finland in 1990. The problem with such interventions is that "doing something" isn't enough. To limit climate change, the world has to reduce net emissions to zero, which requires "doing a lot." The principal challenge facing humanity is how to close the gap identified by Nordhaus and Yang (1996).

Lacking a World Government, the only way in which this gap can be closed is by the development of cooperative institutions; in a word, treaties. The Montreal Protocol on Protection of the Stratospheric Ozone Layer is an example of a treaty that sustains full cooperation. However, just because the world succeeded in overcoming free riding in this one instance doesn't mean that it has the capability to overcome free riding in every instance. Climate change is probably the most difficult collective action problem the world has ever faced.

The Kyoto Protocol, adopted in 1997, a year after Nordhaus and Yang's paper was published, tried to close the gap between the noncooperative and full cooperative outcomes, but failed for lack of an enforcement mechanism. The Paris Agreement, adopted in 2015, abandons Kyoto's pretense that negotiated, "legally binding" emission limits can be strictly enforced and relies instead on national pledges and voluntary compliance, reinforced by "naming and shaming." The Paris pledges were supposed to be chosen so that, when added up, they would achieve the Paris Agreement's collective goal. However, the pledges made in Paris are so modest as to virtually guarantee that temperature will shoot past the collective goal. Making matters worse, there is a reason to think that countries will fall short of their pledges.[5] It is hard to see how Paris can improve much on noncooperation.

The Paris Agreement's collective goal is a partial answer to Nordhaus's first question (partial because it is expressed as a temperature goal rather than as a complete time path). In Paris, nearly all the world's countries agreed that they should act collectively to keep global mean temperature change "well below" 2°C, relative to the pre-industrial level, a goal that was chosen to "significantly reduce the risks and impacts of climate change." In his 1975 paper, long before he estimated an optimal emissions path, Nordhaus (1975, p. 23) reasoned similarly, arguing that "global temperatures more than 2°C or 3°C above the current average temperature" should be avoided, as they "would take the climate outside the range of observations which have been made over the last several hundred thousand years." When Nordhaus expressed this judgement, atmospheric concentrations were at 331 parts per million (ppm). When Paris was adopted, they were at 401 ppm (today, they are over 411 ppm). What was desirable in 1975 was no longer feasible when negotiators met in Paris (unless steps are taken to remove CO_2 from the atmosphere). However, if we interpret the Paris goal as seeking to limit temperature change to 2°C *calculated as a one-hundred-year average*, rather than as a constraint never to be breached in any year, Nordhaus's latest

---

[5]In a lab experiment of the Paris Agreement's design, Astrid Dannenberg and I find that groups of players choose a collective target that is less than the ideal, that individuals in the group make pledges that in aggregate fall short of their target, and that these same players go on to contribute less than they pledged (Barrett and Dannenberg, 2016).

calculations (Nordhaus, 2019a) show that achievement of this goal is not only feasible but gives rise to a temperature path that comes very close to Nordhaus's current projection for the optimal path for global temperature.

The correspondence among these three values — Nordhaus's "first approximation" of a "reasonable standard" or goal for limiting climate change, the collective target agreed in Paris, and Nordhaus's latest calculation of an optimal temperature path — suggests to me that determination of an optimal target and associated time path for emissions, concentrations, and temperature, was never the most critical issue facing humanity. The most critical issue, I have always felt, was, and remains, figuring out how to close the gap between the noncooperative and full cooperative outcomes.

In a recent paper (Nordhaus, 2015), Professor Nordhaus explores a possible remedy to the free riding problem. This involves members of a "climate club" imposing tariffs on nonmembers in the hope that, by doing so, membership in the club would grow and free riding shrink. Ideally, membership would be full, so that trade was unimpeded and free riding vanquished. Nordhaus finds that the ideal scenario only holds if the gains to cooperation on climate change (reflected in the social cost of carbon, a parameter in this model) are small relative to the gains to cooperation on trade. Nordhaus's calculations suggest that linking cooperation on climate change to cooperation on trade is unlikely to be able to close the gap fully.[6] At best, this form of linkage is likely to be no more than a partial remedy.

The quest for a more complete answer to Nordhaus's second question is one that Nordhaus and I, among others, continue to pursue. In my judgement, it is a quest that is likely to fall short of the ideal, and involve a number of measures rather than a single, elegant solution. It is also likely to involve more approaches than just limiting emissions. In particular, it is likely to involve the imperfect and risky solutions that Nordhaus (2015, p. 8) identified in his first paper, such as using "stratospheric dust to cool the earth," an approach we now call "solar geoengineering," and "removing carbon from the air by an industrial process," an approach known today as "carbon geoengineering."

In this introduction, I have emphasized Nordhaus's early papers not only because they were the first papers on the economics of climate change to be written by anyone, but also because they laid the foundation for all future work on this topic. Reading these papers, I am struck mainly by the issues they include and anticipate rather than the ones they neglect and fail to anticipate. However, there is one approach to addressing climate change that, rather surprisingly, Prof. Nordhaus fails to mention in his early papers, perhaps because it is so obvious. This is the approach of adapting to the climate change that is not prevented. In the remainder of this paper, I present a

---

[6]Inspired by Nordhaus (2015), Astrid Dannenberg and I show that, as nonmembers of the club are likely to retaliate by imposing tariffs on members, the conditions that make linkage effective are more restrictive than his paper suggests. We also show that, at best, linkage is a coordination game in which selection of the efficient outcome (cooperation on both climate and trade) is unreliable. If the conditions that make linkage potentially effective hold, our paper suggests that attempts to link should be made multilaterally rather than unilaterally (Barrett and Dannenberg, 2020).

simple model of international cooperation to limit climate change when adaptation is incorporated as an explicit option.[7]

2. On the Relationship Between Abatement and Adaptation

It is often suggested that abatement and adaptation are complements.[8] Certainly, countries will need to adapt; and they should reduce their emissions. However, this observation misses a more fundamental aspect of the relationship. As the parties to the Paris Agreement recognize, though "the current need for adaptation is significant. . . . greater levels of mitigation can reduce the need for additional adaptation."[9] Flipped around, it must also be true that, should adaptation prove easier and more effective, the returns to reducing emissions will be lower. From this perspective, abatement and adaptation are substitutes.

But they are only *imperfect* substitutes. Adaptation can reduce climate change damages; it cannot eliminate them. Though abatement can reduce the need to adapt, because atmospheric concentrations accumulate in the atmosphere, with a proportion of the "extra" CO_2 remaining in the atmosphere for millennia, even very high levels of abatement cannot prevent the climate from changing. Still, countries must strike a balance between these options. The choice of how much to abate is intimately linked to the choice of how much to adapt.[10]

How these options come to be balanced will depend on the institutional arrangements. Abatement is a global public good, whereas adaptation involves the supply of private and local public goods. So long as governance at the local level can be relied upon to respond to public need, and to finance the supply of local public goods through taxation, adaptation will be reasonably efficient, given the extent of climate change. By contrast, unless free rider incentives are overcome, abatement will fall short of, and climate change exceed, the ideal. Relative to the optimal outcome, countries will abate too little and, as a consequence, adapt too much. This raises the question: Does the ability to adapt also affect the ability of a climate treaty to limit free riding?

My model assumes that countries face two binary choices: to abate their emissions or not abate them; to adapt or not to adapt. Metaphorically, countries choose whether to

---

[7]Adaptation is incorporated implicitly if the benefit of abatement is interpreted as the sum of climate change damage and adaptation costs avoided by abatement.

[8]For example, Easterling *et al.* (2004, p. 2) say that, "Adaptation actions and strategies present a complementary approach to those of greenhouse gas mitigation." What they seem to mean, however, is that countries will have to mitigate and adapt to climate change, not that doing more of one will increase the returns to doing more of the other. Similarly, Willbanks *et al.* (2003, p. 34) reason that, "mitigation and adaptation are not necessarily competitors in climate change impact response strategies. In many cases, they are instead complements. If mitigation is successful in keeping climate change impacts to a moderate level, then adaptation can handle a larger share of the resulting impacts." This statement also does not imply that, as more mitigation is undertaken, the returns to adaptation will increase.

[9]Article 7, paragraph 2.

[10]This statement is correct so long as the cross partials in the payoff function are nonzero. If adaptation and mitigation are substitutes, the cross partials will be negative.

replace fossil fuels with windmills and whether to build dikes to hold back rising seas. Section 3 develops the basic model. In Sec. 4, I solve the model for the benchmark cases of noncooperation and full cooperation. Assuming that countries are symmetric, I restrict parameter values such that, in the noncooperative outcome, every country adapts and none abates; and, in the full cooperative outcome, every country abates and none adapts. I could easily admit other possibilities, but this arrangement offers a particularly stark view of two possible futures: a world of windmills or a world of dikes. Section 5 introduces the possibility that countries might negotiate a treaty to limit emissions. Of course, participation in a treaty may be incomplete. I show here that, depending on parameter values, the level of abatement supported by a self-enforcing treaty may be so great as to render adaptation unattractive, or so puny as to make all countries want to adapt anyway. Unfortunately, on its current trajectory, the world seems destined for this second outcome.

3. The Model

There are N symmetric countries, each of which must make two binary choices: whether or not to abate its emissions and whether or not to adapt. Letting q_i denote country i's abatement choice and a_i's adaptation choice, i's problem is to choose $q_i, a_i \in \{0, 1\}$ so as to maximize

$$\pi_i(q_i, a_i; Q_{-i}) = bQ(1 - \alpha a_i) + \alpha a_i b\bar{Q} - cq_i - \gamma a_i, \tag{1}$$

where $Q = q_i + Q_{-i}$, $Q_{-i} = \sum_{j=1, j \neq i}^{N} q_j$, $\alpha \in [0, 1]$, and $b, c, \gamma > 0$.

If nothing is done to reduce emissions, atmospheric concentrations will equal $\bar{Q}$ (the "no control" concentration level), and every country i will suffer damages (a loss) equal to $-b\bar{Q}$. Abatement lowers these damages (in absolute value) to $-b(\bar{Q} - Q)$. For damages to remain a loss even when aggregate abatement is maximal, we require $\bar{Q} > N$. Adaptation by i further reduces these damages (in absolute value) to $-b(\bar{Q} - Q)(1 - \alpha a_i)$ Benefits to i equal the damages avoided by policy — that is, actual damages minus "no control" damages or $-b(\bar{Q} - Q)(1 - \alpha a_i) - (-b\bar{Q})$. This gives the left-hand term in (1). The remaining terms subtract the costs of abatement and adaptation by i.

The parameter α represents the effectiveness of adaptation. If $\alpha = 0$, adaptation is wholly ineffective. If $\alpha = 1$, adaptation is so effective that the marginal benefit of abatement equals zero — for those countries that adapt. All else being equal, the larger is α the more every country i will want to adapt and not to abate. The smaller is α the more every country i will want to abate and not to adapt. For intermediate values, of course, countries may want to do both, to abate and to adapt. The parameter $\bar{Q}$ is also important. If $\bar{Q}$ is very large, it will pay every country to adapt irrespective of the level of global abatement.

Plugging in admissible values for abatement and adaptation gives the following:

$$\pi_i(0, 0; Q_{-i}) = bQ_{-i}, \tag{2a}$$

$$\pi_i(1, 0; Q_{-i}) = b(Q_{-i} + 1) - c, \tag{2b}$$

$$\pi_i(0, 1; Q_{-i}) = bQ_{-i}(1 - \alpha) + \alpha b\bar{Q} - \gamma, \tag{2c}$$

$$\pi_i(1, 1; Q_{-i}) = b(Q_{-i} + 1)(1 - \alpha) + \alpha b\bar{Q} - c - \gamma. \tag{2d}$$

Lemma 1. *Assume*

$$c > b. \tag{3}$$

Then $\pi_i(0, a_i; Q_{-i}) > \pi(1, a_i; Q_{-i}) \forall a_i, Q_{-i}$.

The proof is found by substituting these values in Eqs. (2a)–(2d). No matter how much other countries abate, and whether i adapts or not, i can only lose by abating its emissions.

While each country's incentive to abate does not depend on whether other countries play abate, each country's decision to adapt will depend on the abatement decisions of all countries.

Lemma 2. *Assume*

$$N > \frac{\alpha b\bar{Q} - \gamma}{\alpha b} > 0. \tag{4}$$

Then

$$\pi_i(q_i, 0; Q_{-i}) > \pi_i(q_i, 1; Q_{-i}) \quad \text{for } Q > \hat{Q}, \tag{5a}$$

$$\pi_i(q_i, 0; Q_{-i}) < \pi_i(q_i, 1; Q_{-i}) \quad \text{for } Q < \hat{Q}, \tag{5b}$$

where

$$\hat{Q} = \frac{\alpha b\bar{Q} - \gamma}{\alpha b}. \tag{6}$$

Again, the proof is found by substituting (2a)–(2d) into (5a) and (5b). Intuitively, if the inequality on the left side of (4) holds, then it pays each country not to adapt, given that every country abates; if the inequality on the right side of (4) holds, then it pays each country to adapt, given that no country abates. This implies that there exists a threshold level of global abatement such that, for abatement below this level, every country will adapt; and for abatement above this level, no country will adapt. This threshold is given in (6).

The existence of this threshold raises the question of whether, if enough other countries play Abate, it will pay i to play Abate, so as to render its own adaptation uneconomic. The following lemma establishes that, under the assumptions made thus far, this situation cannot arise.

Lemma 3. *Assume* (3) *and* (4) *and let* $\tilde{Q}_{-i}$ *satisfy* $\hat{Q} > \tilde{Q}_{-i} > \hat{Q} - 1$, *where* $\hat{Q}$ *is given by* (6). *Then*

$$\pi_i(0, 1; \tilde{Q}_{-i}) > \pi_i(1, 0; \tilde{Q}_{-i}). \tag{7}$$

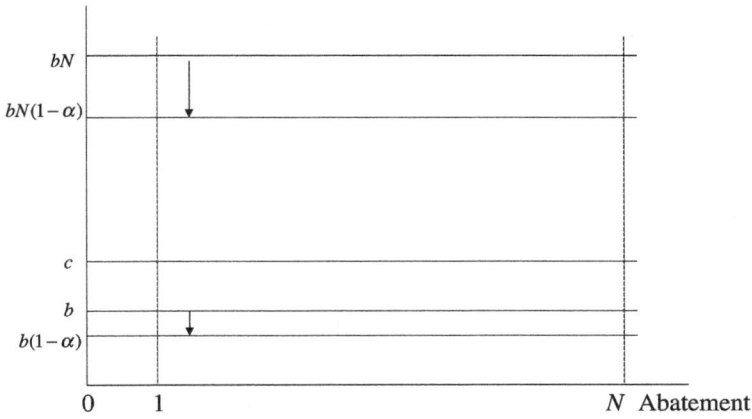

Figure 1. The incentives to abate.

The proof is by contradiction. Suppose (7) does not hold. Then $\pi_i(0, 1; \tilde{Q}_{-i}) \le \pi_i(1, 0; \tilde{Q}_{-i})$ Substituting (2b) and (2c) into this inequality and rearranging gives

$$\tilde{Q}_{-i} \ge \frac{(\alpha b \bar{Q} - \gamma) + (c - b)}{\alpha b}. \tag{8}$$

But by assumption (3), the RHS of this inequality is greater than $\hat{Q}$ and by the definition of $\tilde{Q}_{-i}$, $\hat{Q} > \tilde{Q}_{-i}$. Hence, inequality (7) must hold. Finally, if "switching" abatement for adaptation is not optimal in the neighborhood of the threshold, then for other values of Q_{-i} "switching" will certainly not be optimal.

Figures 1 and 2 (not drawn to scale) illustrate the model. Figure 1 shows how adaptation by every country affects the incentives to abate, both individually and collectively. With or without adaptation, countries do not have an individual incentive to abate. Adaptation lowers the aggregate gains to abatement, but even if every country adapts, the figure shows that all countries collectively are better off if every country

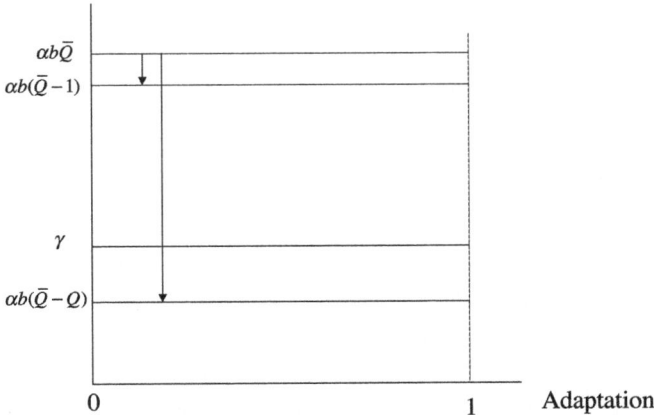

Figure 2. The incentives to adapt.

abates. Figure 2 shows how abatement affects the incentives for countries to adapt. In the figure, abatement by just one country has no impact on this decision (it does, however, reduce the returns to adaptation). Abatement by "enough" countries, however, renders adaptation by every country uneconomic.

4. Preliminary Analysis

The first main result:

Proposition 1. *Assume* (3) *and* (4). *Then, in the noncooperative outcome, every country adapts and none abates.*

From Lemma 1, we know that irrespective of what other countries do, each country is better off not abating. From Lemma 2, we know that, given that no country abates, every country is better off adapting. In equilibrium, every country has a dike and none has a windmill.

Full cooperation requires choosing $q_i, a_i \in \{0, 1\}$ for every i so as to maximize

$$\Pi = \sum_{i=1}^{N} \pi_i(q_i, a_i; Q_{-i}) = bQ(1 - \alpha a_i)N + \alpha a_i b\bar{Q}N - cQ - \gamma a_i N. \tag{9}$$

The solution requires

$$q^{FC} = 0 \quad \text{for } a > \frac{bN - c}{\alpha bN}, \quad q^{FC} = 1 \quad \text{for } a < \frac{bN - c}{\alpha bN}; \tag{10}$$

$$q^{FC} = 0 \quad \text{for } q > \frac{\alpha b\bar{Q} - \gamma}{\alpha bN}, \quad a^{FC} = 1 \quad \text{for } q < \frac{\alpha b\bar{Q} - \gamma}{\alpha bN}. \tag{11}$$

There are two possibilities. Either every country adapts and none abates or every country abates and none adapts. Substituting these values into (9), we find that all countries strictly prefer the latter outcome to the former if and only if the following assumption holds:

$$(bN - c) > (\alpha b\bar{Q} - \gamma). \tag{12}$$

(Note that, by (4), the term on the right-hand side of (12) is strictly positive.) We then have

Proposition 2. *Assume* (12). *Then, in the full cooperative outcome, every country abates and none adapts.*

To sum up, the simple model presented here displays a stark contrast. If countries make their choices independently, every country has dikes and none has windmills. If countries cooperate fully, every country has windmills and none has dikes. These are profoundly different worlds.

Figure 3 illustrates these outcomes. The payoff curves show what each country gets by abating or not abating, taking as given the behavior of other countries.

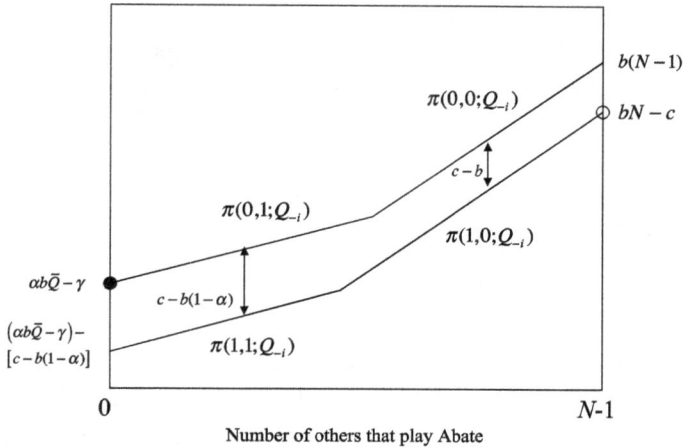

Figure 3. The Abate-Adapt game.

The noncooperative outcome is represented by the closed dot (on the far left of the figure). The full cooperative outcome is depicted by the open dot (far right). Since the payoff curves never cross, not abating is a dominant strategy (Lemma 1). Whether countries will adapt depends on the aggregate level of abatement (Lemma 2). This threshold level of abatement occurs where the two payoff curves are kinked. To the left of this threshold, adaptation is optimal for every country. To the right, each country does better by not adapting. So, in the noncooperative outcome (again, on the left), countries adapt and do not abate. In the full cooperative outcome (right), they abate and do not adapt.

Figure 4 provides a close-up view of the region of the above graph in which the payoff curves are kinked. In this figure, $\hat{k}$ denotes the smallest integer greater than $\hat{Q}$, defined in Eq. (6). Along the bottom payoff curve, the representative country abates.

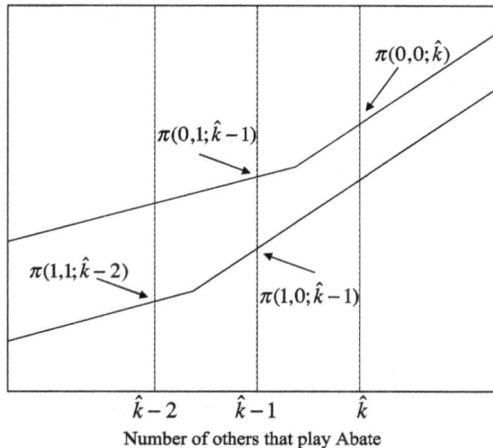

Figure 4. Close up of the Abate-Adapt game.

To the left of the kink, this country adapts; to the right it does not adapt. At $\hat{k} - 1$, $Q = \hat{k} > \hat{Q}$ and all the $\hat{k}$ countries that abate do not adapt; they each get the payoff $\pi(1, 0; \hat{k} - 1)$. Along the top payoff curve, the representative country does not abate. At $\hat{k} - 1$, $Q = \hat{k} - 1 < \hat{Q}$ and all the $N - (\hat{k} - 1)$ countries that do not abate get the payoff $\pi(0, 1; \hat{k} - 1)$.

As Fig. 3 shows, the Nash equilibrium is inefficient. Could a treaty sustain a more efficient outcome? I turn to this question next.

5. Treaty Equilibria

Suppose that some countries cooperate to reduce emissions by means of a treaty. Using the approach described in Barrett (2005), the abatement game consists of three stages. In stage 1, every country chooses independently whether to be a signatory or a nonsignatory to the treaty. In stage 2, signatories choose their abatement levels collectively. In stage 3, nonsignatories choose their abatement levels independently. Here we need to modify this model slightly to accommodate the decision to adapt. This is a unilateral decision, and so is incorporated in a final (fourth) stage.

Stages 3 and 4 are easily solved. From (2), we know that every country will play Adapt if $Q < \hat{Q}$; otherwise, every country will not adapt. Solving stage 3, we know that nonsignatories will never play Abate.

What will signatories do in Stage 2? Signatories can anticipate that their choice of whether to abate will not affect how nonsignatories play in Stage 3. However, they can also anticipate how their choice of whether to abate will affect the choice by every country to adapt. Let there be k signatories. Then, each signatory gets one of three possible payoffs:

$$\pi_s(0, 1; 0) = \alpha b \bar{Q} - \gamma, \tag{13}$$

$$\pi_s(1, 1; k - 1) = bk(1 - \alpha) + \alpha b \bar{Q} - c - \gamma \quad \text{for } k < \hat{Q}, \tag{14}$$

$$\pi_s(1, 0; k - 1) = bk - c \quad \text{for } k > \hat{Q}. \tag{15}$$

Using (13), (14), and (6), signatories will play Abate/Adapt if

$$k > \frac{c}{b(1 - \alpha)} \quad \text{for } k < \frac{(\alpha b \bar{Q} - \gamma)}{\alpha b}. \tag{16}$$

Using (13), (15), and (6), they will play Abate/Don't Adapt if

$$k > \frac{(\alpha b \bar{Q} - \gamma + c)}{b} \quad \text{for } k > \frac{(\alpha b \bar{Q} - \gamma)}{\alpha b}. \tag{17}$$

If neither (16) nor (17) hold, signatories will play Don't Abate/Adapt.

Stage 2 determines the basic obligations established by the treaty. It assumes that the treaty is "collectively rational" in the sense that, for k given, signatories cannot do better collectively except by behaving as specified above.

Stage 1 determines the level of participation. Assuming that signatories and non-signatories behave as calculated in stages 2 and 3, when will a country choose to be a signatory rather than a nonsignatory? The equilibrium participation level must satisfy

$$\pi_s(k^*) > \pi_n(k^* - 1) \quad \text{and} \quad \pi_n(k^*) > \pi_s(k^* + 1). \tag{18}$$

Condition (18) says that k^* is an equilibrium participation level if, at k^*, a non-signatory cannot gain by acceding to the agreement and a signatory cannot gain by withdrawing from it.

Clearly, at k^* it must be collectively rational for signatories to play abate. Otherwise they would have nothing to gain by cooperating. Hence, at k^*, either (16) or (17) must be satisfied.

Consider first the case where signatories play Abate/Adapt. Let $\check{k}$ denote the smallest integer that satisfies the first constraint in (16). Then (18) requires

$$\pi_s(1, 1; \check{k} - 1) > \pi_n(0, 1; 0) \quad \text{and} \quad \pi_n(0, 1; \check{k}) > \pi_s(1, 1; \check{k}). \tag{19}$$

After substituting and reducing, conditions (18) are satisfied if

$$\check{k} > \frac{c}{b(1 - \alpha)} \quad \text{and} \quad c > b(1 - \alpha). \tag{20}$$

The first condition is true by the definition of $\check{k}$. The second condition in (20) is satisfied by (3) (given that $\alpha \in [0, 1]$). $\check{k}$ is an equilibrium provided both conditions in (16) are satisfied. These conditions reduce to

$$\frac{\alpha c}{(1 - \alpha)} < \alpha b \bar{Q} - \gamma. \tag{21}$$

For smaller participation levels, abatement is never collectively optimal and an equilibrium agreement therefore does not exist. What about for $k > \check{k}$? Suppose that there are $\check{k} + 1$ signatories. Then, should one of these countries withdraw, the withdrawing country will get a payoff of $b\check{k}(1 - \alpha) + \alpha b \bar{Q} - \gamma$. If the country stays in the agreement, it will get $b(\check{k} + 1)(1 - \alpha) + \alpha b \bar{Q} - c - \gamma$. The former payoff exceeds the latter if $c > b(1 - \alpha)$, which holds by (3). Hence, such a participation level cannot be self-enforcing. Applying the same calculation to higher participation levels, it is easy to show that, when signatories play Abate and Adapt, no agreement consisting of more than $\check{k}$ countries is self-enforcing. The equilibrium is thus $k^* = \check{k}$.

Now let $\widehat{k}$ denote the smallest integer satisfying the inequality on the LHS of (17). In this case, signatories play abate/not adapt and (18) requires

$$\pi_s(1, 0; \widehat{k} - 1) > \pi_n(0, 1; 0) \quad \text{and} \quad \pi_n(0, 0; \widehat{k}) > \pi_s(1, 0; \widehat{k}). \tag{22}$$

Substituting and reducing gives

$$\widehat{k} > \frac{\alpha b \bar{Q} - \gamma + c}{b} \quad \text{and} \quad c > b. \tag{23}$$

The second condition in (23) is satisfied by (3). The first condition is satisfied by the definition of $\hat{k}$. $\hat{k}$ is an equilibrium provided both conditions in (17) are satisfied. These conditions are sure to apply provided

$$\frac{\alpha}{(1-\alpha)} > \alpha b \bar{Q} - \gamma. \tag{24}$$

Again, for smaller participation levels, abatement is never collectively optimal and an equilibrium agreement therefore does not exist. It is also easy to show that an agreement consisting of more than $\hat{k}$ countries cannot be self-enforcing. The unique equilibrium is thus $k^* = \hat{k}$.

The table and proposition below summarize the results:

Proposition 3. *Depending on parameter values, the self-enforcing abatement treaty sustains an outcome in which either all countries adapt or none adapts.*

Intuitively, a higher value for $\bar{Q}$ and a lower value for γ both increase the returns to adaptation, making it more likely that the condition in the top row of Table 1 will hold. A higher value for b also increases the returns to adaptation. In addition, a higher value for b causes participation, and thus abatement, to fall, reinforcing the incentive to adapt. A higher value for c doesn't affect the returns to adaptation directly, but by increasing the participation level and thus the level of abatement, it lowers the returns to adaptation indirectly. The effect of changes in α are more complicated to analyze. On one hand, a higher value for α increases the returns to adaptation directly. On the other hand, a higher value for α decreases the returns to abatement provided countries choose to adapt; and in doing so causes participation to increase, thereby increasing abatement and thus reducing the returns to adaptation indirectly.

If adaptation is infeasible, then k^* will equal the smallest integer greater than c/b (Barrett, 2005). However, from Table 1, we know that, if adaptation is feasible, then the participation level of the self-enforcing IEA must be at least as large as this and may be substantially greater. Hence, we also have

Proposition 4. *The feasibility of adaptation never lowers the participation level of the self-enforcing abatement treaty. It may, however, increase the participation level, irrespective of whether states actually adapt in equilibrium.*

The reason for this result is that, in this linear model, adaptation lowers the marginal benefit of abatement, meaning that more countries must participate in order for

Table 1. Equilibria for the climate treaty game.

If the following condition holds	then k^* is the smallest integer greater than	and in equilibrium, all countries will play
$(\alpha b \bar{Q} - \gamma) < \frac{\alpha c}{(1-\alpha)}$	$\frac{c}{b(1-\alpha)}$	Adapt
$(\alpha b \bar{Q} - \gamma) < \frac{\alpha c}{(1-\alpha)}$	$\frac{\alpha b \bar{Q} - \gamma + c}{b}$	Don't Adapt

abatement to be collectively rational. Even though abatement and adaptation are substitutes in this model, the possibility of adaptation increases, rather than decreases, cooperation as measured by the treaty participation level. In other respects, however, the equilibrium has the familiar property that, for large N, signatories benefit only marginally from the IEA relative to the noncooperative outcome. Nonsignatories — the free riders — benefit more. Overall, however, for large N, a treaty helps very little.

6. Final Comments

The model presented here is simple. The assumption that both abatement and adaptation are binary choices gives stark results. An obvious extension would be to define abatement and adaptation as continuous variables. Different functional forms could also be explored. Of particular concern, however, is the assumption that countries are symmetric. Nordhaus's work has always leaned towards empirical realism, and thus emphasized asymmetries. Once asymmetries are incorporated, concerns of fairness loom large. In the model with symmetric countries, the most obvious concern as regards fairness is free riding. In a model with asymmetric countries, additional concerns arise. Should the countries that contributed most to the historical build-up of $\bar{Q}$ compensate the countries that contributed least for their abatement or adaptation or both? Should they also compensate the latter countries for the "loss and damage" they experience net of any adaptation? What incentives do the countries with special historical responsibilities have to assist the less developed, more vulnerable countries? Addressing climate change fundamentally by reducing net emissions to zero worldwide requires the participation of all countries, and a collective effort that is without precedent. Fairness issues will need to be addressed. But enforcing cooperation, the focus of my model, is the most critical requirement, for without it the climate will continue to change, possibly in ways that are catastrophic for humanity, and there will be no collective gain for the world to share.

The Nobel committee recognized Nordhaus's efforts to answer his first question, but addressing climate change also requires answers to his second question. Finding answers to this second question — in particular, devising ways to sustain collective action in limiting emissions — should be our focus now.

Acknowledgment

I am grateful to Reyer Gerlagh and participants at the Environment, Technology, and Uncertainty Workshop held in Oslo, 23 May 2008 for comments on an early version of the model presented in this paper.

References

Barrage, L (2019). The Nobel memorial prize for William D. Nordhaus. *Scandinavian Journal of Economics* 121(3), 884–924.

Barrett, S (2005). *Environment and Statecraft: The Strategy of Environmental Treaty-Making*. Oxford: Oxford University Press (paperback edition).

Barrett, S and A Dannenberg (2016). An experimental investigation into 'Pledge and Review' in climate negotiations. *Climatic Change* 138(1), 339–352.

Barrett, S and A Dannenberg (2020). The promise and peril of linking trade agreements to the supply of global public goods, Columbia University and University of Kassel.

Easterling, WE, BH Hurd and JB Smith (2004). Coping with global climate change: The role of adaptation in the United States, Pew Center on Global Climate Change.

Nordhaus, WD (1975). Can We Control Carbon Dioxide?. IIASA Working Paper 75–63.

Nordhaus, WD (1977a). Strategies for the Control of Carbon Dioxide. Cowles Foundation Discussion Paper No. 443.

Nordhaus, WD (1977b). Economic growth and climate: The carbon dioxide problem. *American Economic Review Papers and Proceedings* 67(1), 341–346.

Nordhaus, W (2015). Climate clubs: Overcoming free-riding in international climate policy. *American Economic Review* 105(4), 1339–1370.

Nordhaus, WD (2017). Revisiting the social cost of carbon. *Proceedings of the National Academy of Sciences* 114(7), 1518–1523.

Nordhaus, W (2019a). Climate change: The ultimate challenge for economists. *American Economic Review* 109(6), 1991–2014.

Nordhaus, W (2019b). Can we control carbon dioxide? (from 1975). *American Economic Review* 109(6), 2015–2035.

Nordhaus, WD and Z Yang (1996). A regional dynamic general-equilibrium model of alternative climate-change strategies, *American Economic Review* 86(4), 741–765.

CHAPTER 6

SELFISH BUREAUCRATS AND POLICY HETEROGENEITY IN NORDHAUS' DICE

RICHARD S. J. TOL

Department of Economics, University of Sussex
Jubilee Building, Falmer, BN1 9SL, UK

Institute for Environmental Studies
Vrije Universiteit, Amsterdam, The Netherland
Department of Spatial Economics
Vrije Universiteit, Amsterdam, The Netherlands

Tinbergen Institute, Amsterdam, The Netherlands

CESifo, Munich, Germany

Payne Institute for Earth Resources
Colorado School of Mines, Golden, CO, USA
r.tol@sussex.ac.uk

Nordhaus' seminal DICE model assesses first-best climate policy, a useful but unrealistic yardstick. I propose a measure of policy inefficacy if carbon prices are heterogeneous and use observed prices to recalibrate the DICE model. I introduce a Niskanen-inspired model of climate policy with selfish bureaucrats, and calibrate it to carbon dioxide emissions in the European Union and the policy models used by the IPCC. This model also implies a measure of policy inefficacy that I use to recalibrate DICE. The optimal global mean temperature is 1°C perhaps 2°C higher in the recalibrated than in the original DICE model.

Keywords: Climate policy; price heterogeneity; selfish bureaucrats.

1. Introduction

Professor Dr. William D. Nordhaus was the first economist to discuss climate change, wondering whether we can control carbon dioxide (Nordhaus, 1975). Nordhaus (1977) concluded that we can, presenting a way to do so at the lowest possible cost. d'Arge (1979) showed that the impacts of climate change — nuclear winter in his case — can be monetized. Building on that, Nordhaus (1982) first analyzed statically optimal climate policy, an analysis he refined later with newer numbers (Nordhaus, 1991a,b), with dynamic optimality (Nordhaus, 1992, 1993), and again with multiple countries (Nordhaus and Yang, 1996). Nordhaus thus showed the economics profession how to

This chapter was originally published in Climate Change Economics, Vol. 11, No. 4, December 2020, published by World Scientific Publishing, Singapore. Reprinted with permission.

analyze climate policy. Although climate change is the mother of all externalities — global, ubiquitous, long-term, uncertain, inequitable — it is not beyond economic analysis, as demonstrated by Nordhaus' work.

Throughout his career, Nordhaus has been firmly committed to analyzing first-best climate policy, which is indeed the yardstick against which to measure any and all climate policy. However, actions by governments, taken in the name of climate change, are sometimes easier to understand as exercises in rent-seeking and rent-creation than as exercises in greenhouse gas emission reduction. Unfortunately, the climate economics literature has largely followed in Nordhaus' footsteps, assuming smart, well-informed, and selfless social planners. In this paper, I set out two alternatives — selfish bureaucrats and heterogeneous carbon prices — and explore the implications using Nordhaus' seminal DICE model. This allows us to explore the distance between Nordhaus' ideal world with its first-best policy and the actual world with its crummy third-best policies. Policy analysts should, of course, strive to move the actual world closer to the ideal one.

Pigou (1920) argued for the use of taxes and subsidies to internalize externalities and restore efficiency (see Bator, 1958, for a formal treatment). Dales (1968) added tradable permits. The desirable properties of these policy instruments are well-established (Baumol and Oates, 1971; Baumol, 1972; Montgomery, 1972). It is also well-known that the equimarginal principle does not hold for emission reduction costs if there is market power (Buchanan, 1969; Barnett, 1980), distortionary taxation (Krutilla, 1991; Goulder *et al.*, 1999; Barrage, 2018, 2020), a second market imperfection (Fischer, 2008), or a concern for equity (Stiglitz, 2019) — carbon taxes should rarely be uniform. Indeed, the double dividend literature designs policy instruments to take advantage of prior market distortions so as to reduce the costs of greenhouse gas emission reduction (Goulder, 1995; Patuelli *et al.*, 2005; van Heerden *et al.*, 2006).

I do not consider such arguments in this paper. Although a uniform carbon price is the cheapest way to reduce greenhouse gas emissions in an economy with a single imperfection, it requires international cooperation on providing a public good, which is hard (Carraro and Siniscalco, 1992; Barrett, 1994; Nordhaus and Yang, 1996). International tax harmonization is beyond current politics; even Member States of the European Union jealously guard their sovereignty over fiscal policy. Linking national systems of tradable permits would be an alternative route to a globally uniform carbon price (Rehdanz and Tol, 2005; Haites, 2016) but has floundered (Ranson and Stavins, 2016). So we observe different carbon prices in different parts of the world. Also within countries, policy implementation is far from perfect. This implies that climate policy is needlessly expensive (Böhringer *et al.*, 2009) or that, for any given expenditure on climate policy, emission reduction is needlessly ineffective. In Sec. 2, I present a measure of policy inefficacy, calibrate it, and adjust DICE to take this into account.

This first calibration is phenomenological. In Sec. 3, I consider a structural model inspired by Niskanen (1971). I introduce unnecessary administrative costs for emissions monitoring and compliance with regulations, and bureaucrats who seek to increase their desk-size by imposing unnecessary costs on emitters. Policy is no longer optimal, but needlessly expensive to serve the needs not of society but rather the climate bureaucracy. I calibrate this model, and extend DICE to include a climatocracy.

To the best of my knowledge, I am the first to marry Nordhaus to Niskanen. I am fairly confident this is the case for the economics of climate change. For environmental economics more broadly, Oates and Strassmann (1978, 1984) show that if bureaucrats pollute, they should be taxed just like other polluters. Lyon (1990) notes that this is not true if pollution abatement affects desk size. None of these papers consider the case studied here, where bureaucrats use emission control to extend their desk.

Section 4 shows the implications for emissions and climate, and Sec. 5 concludes.

2. Heterogeneous Taxes

2.1. *Uniform taxation*

Let us consider a tax τ to incentivize emission reduction, measured by relative emission reduction effort $0 <= R <= 1$. Polluters, indexed by $i = 1, 2, \ldots, I$, have costs C_i:

$$C_i = 0.5\alpha_i R_i^2 Y_i + \tau(1 - R_i)E_i, \tag{1}$$

where E_i are uncontrolled emissions, Y_i is output, and α_i is a parameter. The first-order condition for cost-minimization is

$$\frac{\partial C_i}{\partial R_i} = -\tau E_i + \alpha_i R_i Y_i = 0 \Rightarrow R_i = \tau \frac{E_i}{\alpha_i Y_i}. \tag{2}$$

Emissions are reduced further if the tax is higher, or emission reduction cheaper. Suppose the benefits of emission reduction are given by

$$B = \beta \sum_i R_i E_i. \tag{3}$$

Maximizing net benefits $B - \sum_i 0.5\alpha_i R_i^2 Y_i$ (taxes are transfers) leads to optimal emission reduction $R_i = \beta \frac{E_i}{\alpha_i Y_i}$ and Pigou tax $\tau = \beta$.

2.2. *Heterogeneous taxes*

For a uniform tax τ, the total costs of emission reduction (net of taxes) equal

$$C_U(\tau) = \sum_i C_i = 0.5\tau^2 \sum \frac{E_i^2}{\alpha_i Y_i} \tag{4}$$

and for heterogeneous taxes τ_i

$$C_H(\tau_i) = 0.5 \sum_i \tau_i^2 \frac{E_i^2}{\alpha_i Y_i}. \tag{5}$$

Total emission reduction for a uniform tax is

$$R_U(\tau) = \sum_i R_i E_i = \tau \sum_i \frac{E_i^2}{\alpha_i Y_i} \tag{6}$$

and for heterogeneous taxes

$$R_H(\tau_i) = \sum_i \tau_i \frac{E_i^2}{\alpha_i Y_i}. \tag{7}$$

The two cases can be compared, either by considering the difference in emission reduction for the same cost or by the difference in costs for the same emission reduction. To start with the latter, the same total emission reduction $R_U = R_H$ implies an equivalent uniform tax

$$\tau_R = \frac{\sum_i \tau_i \frac{E_i^2}{\alpha_i Y_i}}{\sum_i \frac{E_i^2}{\alpha_i Y_i}} \tag{8}$$

and relative cost

$$\frac{C_H(\tau_i)}{C_U(\tau_R)} = \frac{\sum_i \frac{E_i^2}{\alpha_i Y_i} \sum_i \tau_i^2 \frac{E_i^2}{\alpha_i Y_i}}{\left(\sum_i \tau_i \frac{E_i^2}{\alpha_i Y_i}\right)} > 1. \tag{9}$$

The inequality follows as a uniform tax is the least-cost solution for a given target. Heterogeneous taxes achieve the same emission reduction at a higher cost. Equation (9) is a measure of the relative *inefficiency* of heterogeneous taxation.

The same total emission reduction costs $C_U = C_H$ also implies an equivalent uniform carbon tax

$$\tau_C = \sqrt{\frac{\sum_i \tau_i^2 \frac{E_i^2}{\alpha_i Y_i}}{\sum_i \frac{E_i^2}{\alpha_i Y_i}}} \tag{10}$$

and relative emission reduction

$$\frac{R_H(\tau_i)}{R_U(\tau_C)} = \frac{\sum_i \tau_i \frac{E_i^2}{\alpha_i Y_i}}{\sqrt{\sum_i \frac{E_i^2}{\alpha_i Y_i} \sum_i \tau_i^2 \frac{E_i^2}{\alpha_i Y_i}}}. \tag{11}$$

We see that

$$\frac{R_H(\tau_i)}{R_C(\tau_C)} = \sqrt{\frac{C_U(\tau_R)}{C_H(\tau_i)}} < 1 \tag{12}$$

so that less emission reduction is achieved for the same total cost. Equation (12) is a measure of the relative *inefficacy* of heterogeneous taxation.

If we assume $\frac{E_i^2}{\alpha_i Y_i} = 1$ then $\tau_R = \sum_i \tau_i$ and

$$\frac{C_H(\tau_i)}{C_U(\tau_R)} = \frac{I \sum_i \tau_i^2}{(\sum_i \tau_i)^2}. \tag{13}$$

This ratio is greater than one, because numerator minus denominator equals the variance of τ_i times I^2. We also have $\tau_C = \sqrt{\sum_i \tau_i^2}$ and

$$\frac{R_H(\tau_i)}{R_U(\tau_C)} = \frac{\sum_i \tau_i}{\sqrt{I \sum_i \tau_i^2}}. \tag{14}$$

The interpretation is intuitive: The greater the spread of carbon prices, the lower the relative efficacy of climate policy. This interpretation carries over to Eq. (12), with moments replaced by weighted moments.

2.3. *Calibration*

Figure 1 shows the histogram of carbon taxes applied in various countries in the world[1] and recent prices of emission permits in cap-and-trade system.[2] Carbon taxes vary widely, from \$0.30/tC in Poland to \$466/tC in Sweden. Carbon prices vary less widely but still show a considerable range, from \$4/tC in Chongqing to \$107/tC in the European Union. These carbon taxes and prices imply that $R_H/R_U = 0.63$ in Eq. (14) — that is, heterogeneity of carbon prices reduces the relative efficacy of climate policy by more than a third.

This measure of price heterogeneity ignores that the majority of countries do not have any carbon price at all. Including these, relative efficacy falls to 25%.

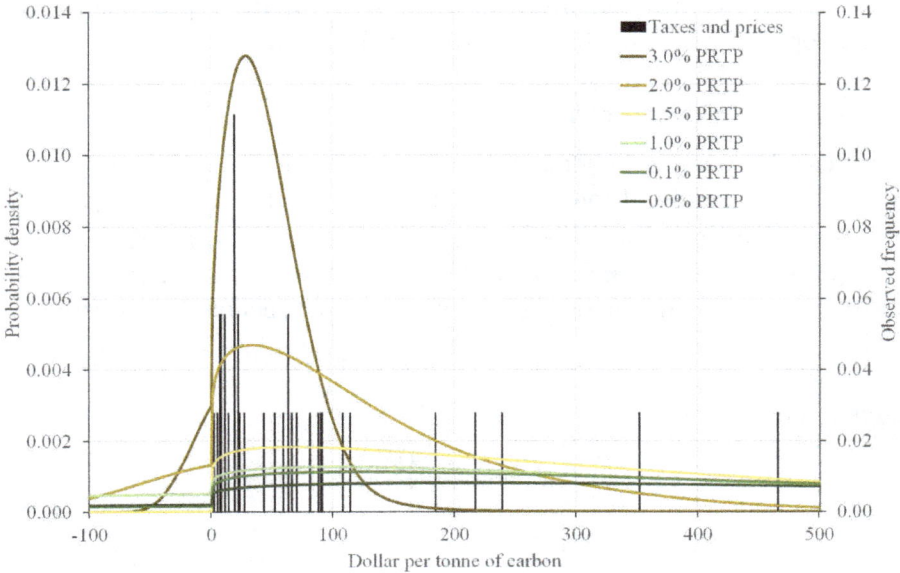

Figure 1. Histogram of observed carbon taxes and prices (bars) and probability density of estimated social cost of carbon for various pure rates of time preferences (lines).

---

[1]Source: World Bank Carbon Pricing Dashboard. https://carbonpricingdashboard.worldbank.org/map_data
[2]Source: International Carbon Action Partnership. https://icapcarbonaction.com/en/ets-prices

Heterogeneous climate policy means that, for the given cost, only a quarter of emission reduction is achieved. This is such an extreme result that I ignored it. Instead, I derive an alternative sensitivity analysis below.

Figure 1 also shows the probability density functions of the published estimates of the social cost of carbon for frequently used pure rates of time preference (data from Tol, 2018). As the observed carbon prices are the same order of magnitude as the estimates of the *global* social cost of carbon, the range of carbon prices is more readily interpreted as heterogeneity rather than as noncooperative behavior (see Ricke *et al.*, 2018; Tol, 2019, for estimates of the *national* social costs of carbon).

Comparing the histogram of the observed carbon prices to the probability densities of the estimated social cost of carbon, current climate policy appears to be based on relatively high utility discount rates. The *Mean Integrated Squared Error* is 3.9% for a pure rate of time preference of 3.0%, 4.0% for 2.0%, 4.2% for 1.5%, and 4.3% for 1.0% or less. This vindicates Nordhaus' argument that the discount rate used for climate policy analysis should be close to the discount rate used to evaluate other investments (Nordhaus, 1997, 2007).

3. Emission Reduction with Bureaucrats

3.1. *Zero administrative cost*

Equation (1) assumes that administrative costs are zero, and Eq. (2) shows the corresponding optimal response of a polluter to a pollution tax.

3.2. *Variable administrative cost*

Now suppose that there is an administrative cost AR_iE_i for reporting, proportional to emissions *reduced*. Firms already monitor and report energy use, so monitoring and reporting emissions is virtually costless — emissions equal energy use by fuel times the fuel-specific emission coefficient. The assumption here is that there is additional, unnecessary reporting on emission reduction. The cost function then becomes

$$C_i = 0.5\alpha_i R_i^2 Y_i + AR_i E_i + \tau(1 - R_i)E_i \tag{15}$$

and emission reduction

$$R_i = \frac{(\tau - A)E_i}{\alpha_i Y_i}. \tag{16}$$

That is, variable administrative costs reduce abatement. If R_i is capped from below at 0, variable administrative costs may even prevent abatement altogether if the marginal administrative cost exceeds the tax rate.

Equation (15) could be extended further with a fixed cost for administration. This would not affect the first-order conditions and hence firm behavior. Firms that do not reduce their emissions would still face administrative costs. However, the social planner may abstain from imposing a carbon tax if the net benefits of doing so are smaller than the fixed administrative costs.

3.3. *Selfish administrators*

Now introduce a regulator who aims to maximize net social benefits as well as the size of the administration. That is, the social planner is no longer a strictly benevolent social planner, but rather one who acts with a mix of self-interest and social interest, in the spirit of Niskanen (1971). One way to formulate this is

$$W = \beta \sum_i R_i E_i - \sum_i 0.5\alpha_i R_i^2 Y_i - A \sum_i R_i E_i + \phi \ln\left(A \sum_i R_i E_i\right). \quad (17)$$

The first-order conditions are

$$\frac{\partial W}{\partial A} = \frac{\phi}{A} - \sum_i R_i E_i = 0 \Rightarrow A = \frac{\phi}{\sum_i R_i E_i} \quad (18)$$

and, assuming $\tau > A$,

$$\frac{\partial W}{\partial \tau} = \left(\beta - A + \frac{\phi}{\sum_i R_i E_i}\right) \sum_i \frac{E_i^2}{\alpha_i Y_i} - \sum_i \alpha_i Y_i \frac{(\tau - A)E_i^2}{\alpha_i^2 Y_i^2} = 0 \Rightarrow \tau = \beta - A. \quad (19)$$

This simplifies to the Pigou tax for $\phi = 0$.

There are two other interesting special cases. First, if $\beta = 0$ — anthropogenic climate change is not real or climate change does no damage — then the optimal carbon tax is negative, that is, a subsidy.[3] In this case, bureaucrats would still want to meddle with emissions but have no reason to prefer emission reductions over emission increases. Because τ and A have opposite signs in Equation (16), a carbon subsidy is a more effective way to raise administrative revenue.

Second, if bureaucrats are given a free hand and do not care about the costs imposed on society then, in this static model, they would maximize the size of the administration by setting $R_i = 1 \forall i$.

3.4. *Calibration*

In the model, the carbon tax is split into two: One part goes to reducing emissions, and another part to financing a climate bureaucracy. This implies that a carbon tax is less effective in reducing emissions than it could have been. As there are no observations of the size of the climatocracy, the model can only be calibrated by contrasting actual greenhouse gas emission reduction to expected greenhouse gas emission reduction.

This is problematic in two ways. First, we do not observe actual emission *reduction*. We observe actual emissions, or rather we impute greenhouse gas emissions from

---

[3]This highlights the partial equilibrium nature of the model: The bureaucrats have no interest in or control over the overall size of the government budget.

energy use and agricultural production. Emission reduction is the difference between the actual and a counterfactual, what emissions would have been had there been no climate policy. I solve this as follows. The World Bank has good data, going back to 1960, for carbon dioxide emissions and economic activity for the countries that currently make up the European Union. Between 1960 and 2004, the emission intensity of the European economies fell, on average, by 1.8% per year. Extrapolating that trend (and assuming economic growth as observed), emissions in 2017 would have been 3.8 million metric tonnes of carbon dioxide. Observed emissions were 3.3 $MtCO_2$.

Figure 2 shows the observed emissions and the projected emissions for the 28 countries of the European Union, including the 95% confidence interval of the projection, as well as the observed and projected emission intensity. This simple analysis suggests that climate policy in the EU has statistically significantly reduced emissions.

Since 2005, roughly half of EU emissions have been regulated by a system of cap-and-trade. Using the observed prices and a model for climate policy evaluation tells us how much emissions should have fallen. However, and this is the second problem, models disagree strongly about the impact of climate policy on emissions. This is typically reported as a wide range of carbon prices needed to meet a particular target (Clarke *et al.*, 2014), but that can of course be inverted to a wide range of emission reductions resulting from a particular carbon price. The IIASA database[4] has scenarios with and without climate policy for 19 different models. Regressing emission reduction in 2020 on carbon price, controlling for model fixed effects and clustering standard errors at the model level, I find that emissions fall by 0.3% per $1/tCO2.

Figure 3 shows the results per model. The figure displays the efficacy of near-term climate policy, here measured as the fraction emission reduction from baseline in 2020 divided by the carbon tax in the same year. The models strongly disagree. The most pessimistic model finds that a carbon tax of $1/tCO_2$ would reduce emissions by 0.1%, the most optimistic model by more than 1.5%. The weighted average of 0.3% is closer to the more pessimistic end of the spectrum. I therefore also show the median (of the reported scenarios), which is about 0.6% per $1/tCO2.

If the average of these models is correct, given the observed carbon prices, EU emissions should have fallen to 2.4 $MtCO_2$. See Fig. 2. The "observed" emission reduction is only 30% of the "expected" emission reduction. That is, more than two-thirds of the carbon tax was wasted on, in this interpretation, administrative costs.[5]

The EU ETS applies to only half of greenhouse gas emissions, so inefficacy is more than one-third rather than more than two-thirds. Then again, EU Member States are supposed to regulate the other half of greenhouse gas emissions by means other than the EU ETS. I therefore use an inefficacy of one-third as the base case, and two-thirds

---

[4]See AR5 Scenario Database Explorer.

[5]Other interpretations are possible, of course. Actual emission reduction may fall short of predicted emission reduction because the prediction did not take account of market power, because price volatility was ignored, or because market operators do not expect climate policy to last. The sensitivity analysis below indicates the sensitivity of the key findings to the particular numerical assumptions.

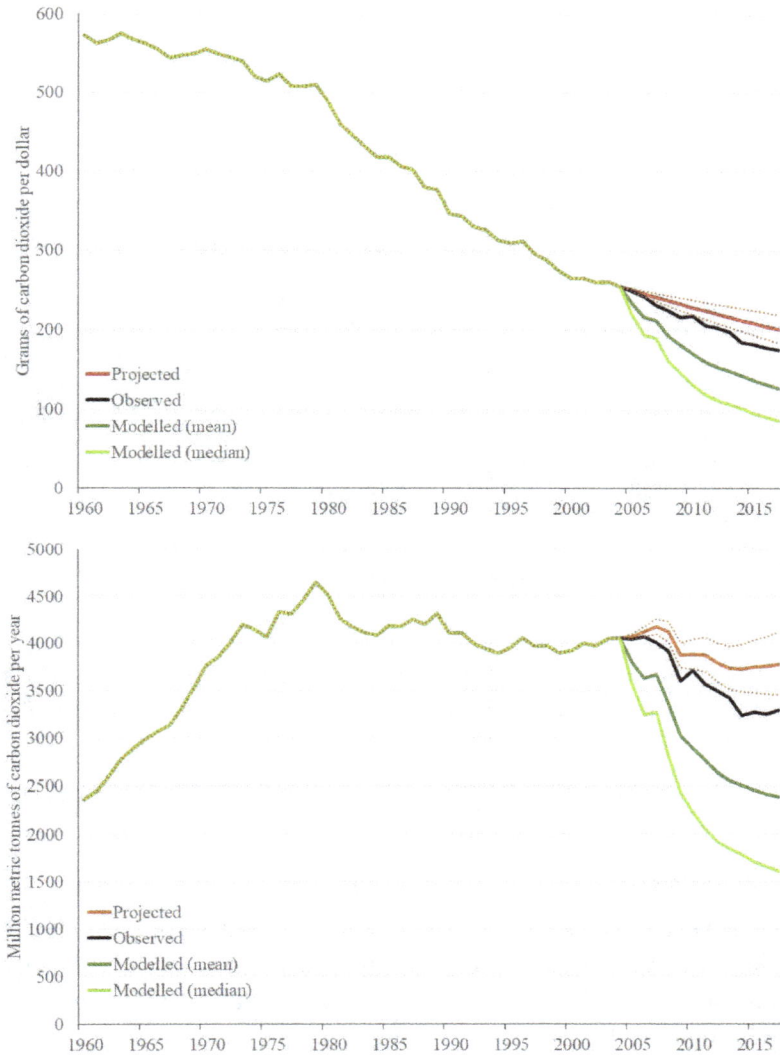

Figure 2. Carbon dioxide emissions (bottom panel) and emission intensity (top panel) in the countries of the European Union as observed, projected, and predicted by models.

in a sensitivity analysis. After all, the EU ETS is not the only policy instrument used to reduce emissions in the economic sectors covered by the EU ETS; there are subsidies for renewables in power generation, for instance, and technological standards for the production of chemicals and steel.

If the median of these model runs is correct, given the observed carbon prices, EU emissions should have fallen to 1.6 $MtCO_2$. See Fig. 2. The "observed" emission reduction is only 17% of the "expected" emission reduction. That is, more than four-fifths of the carbon tax was diverted to administrative costs. This is such an extreme outcome that it is ignored below.

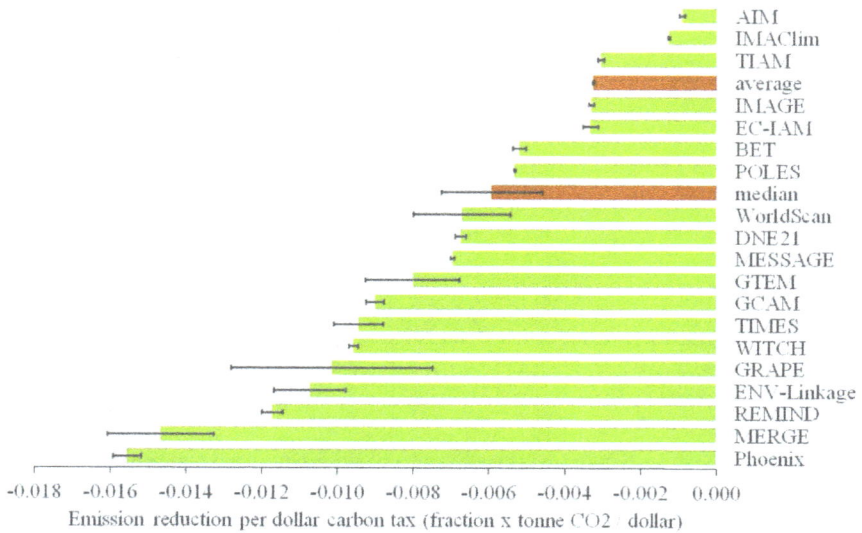

Figure 3. Efficacy of near-term climate policy by model.

Note that the median is far from the optimistic extreme. See Fig. 3. The models in the IIASA database are the ones used in the assessment reports of the IPCC. These models appear to be biased towards cheap climate policy.

4. Results

4.1. *Amending DICE*

I use the Excel version of DICE2016. Amending DICE is straightforward. Following the discussion above, only a fraction (two-thirds) of the marginal damage costs is effective as a carbon tax. This is my base case. As a first sensitivity analysis, I reduce that fraction to one-third.

This simple amendment ignores two complications. The first is that if the carbon tax is less than the Pigou tax, climate change will be more pronounced and the Pigou tax consequently higher. Then again, emission reduction costs are lower, economic growth faster, and the Pigou tax consequently higher. In the implementation used here, the former effect dominates the later prior to 2050. I show results with and without this effect.

The second complication is that the costs of emission reduction are different, and hence economic growth. Specifically, the costs of emission reduction are quadratic in abatement, in both the model above and in DICE, but administrative costs are linear. This means that, for the same carbon tax, total emission reduction costs are lower; average emission reduction costs, per unit of emissions avoided, are higher. I show results with and without this effect.

Administrative costs do not affect GDP. The administrative burden relocates resources from private emitters to the public sector. I ignore the impact of a larger or smaller public sector on the growth prospects of the economy. While a parasitic bureaucracy contributes to GDP, its contribution of welfare is less clear-cut. I therefore only show results below for carbon taxes, emissions and temperatures.

4.2. *Findings*

Figure 4 shows the *effective* carbon price for six scenarios. The carbon price is highest in the original DICE scenario, which represents globally cooperative climate policy, and lowest in the base case, which represent noncooperative climate policy. These two scenarios are due to Nordhaus.

If carbon prices are heterogeneous or used to finance a climatocracy, the signal to reduce emissions is muted. This is shown in the curves labeled "low inefficacy" and "high inefficacy". The size of the effect is as assumed, one-third or two-thirds.

Figure 4 also reveals that the effects of the feedbacks of climate policy on climate change ("Pigou tax") and via abatement costs on economic growth ("administrative costs") are minimal.

The pattern shown in Fig. 4 is reflected in Fig. 5. In the base scenario, in a hundred years' time, emissions are some 85% of what they would have been without any climate policy. In the original, cooperative scenario, the economy is fully decarbonized a century from now — virtue of the assumed backstop. If one-third of climate policy is wasted on price heterogeneity or administrative procedures, full decarbonization is achieved twenty years later. If two-thirds is wasted, emission reduction is only 55% in 2135.

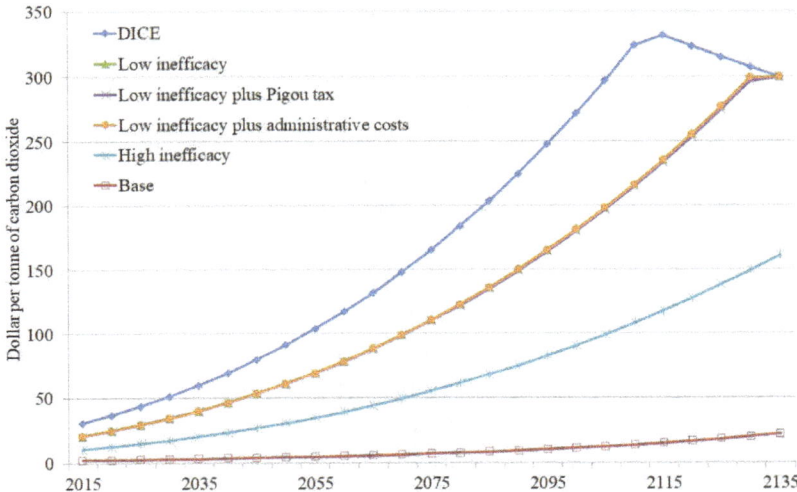

Figure 4. The effective carbon tax for six alternative scenarios.

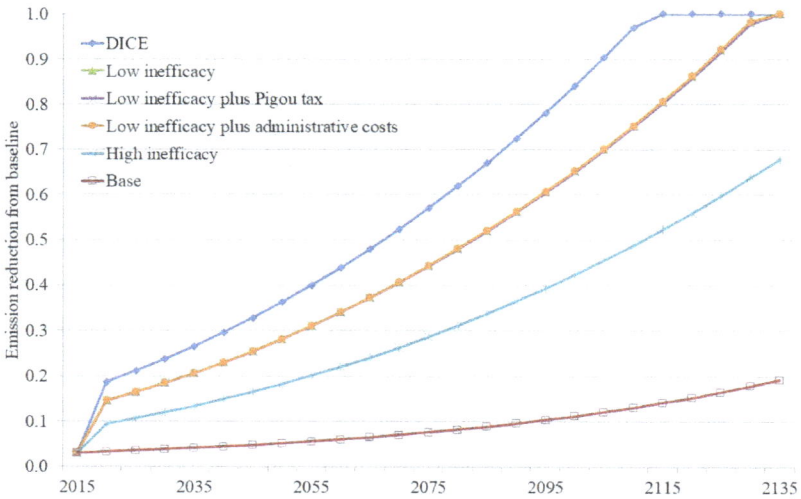

Figure 5. Emission reduction from baseline in six alternative scenarios.

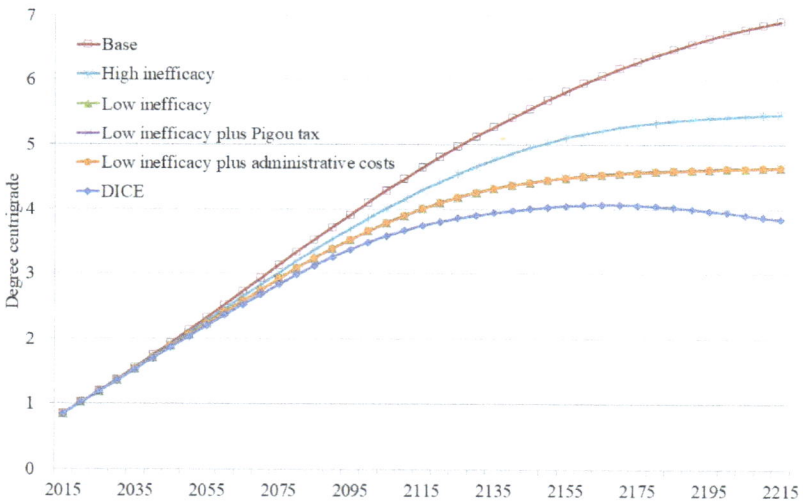

Figure 6. Global annual mean surface air temperature for six alternative scenarios.

Figure 6 shows the same order. In the base case, the world warms by 7°C by the start of the 23rd century, with temperatures still rising. This is reduced to 4°C and falling, in Nordhaus' first-best scenario. Ineffective climate policy adds about 0.75°C warming, and highly ineffective climate policy 1.5°C.

The size of the climatocracy is limited. In the base case, expenditures of climate administration cost peak at 0.22% of total economic output in 2090. After that, the economy grows faster than the carbon price times the emission control rate times emissions. If climate policy is highly ineffective, expenditure on the climatocracy peaks in 2120 at 0.24%. This is a substantial sum of money to waste, but can conceivably escape public and political scrutiny.

5. Conclusion

Professor William D. Nordhaus taught the economics profession how to analyze first-best climate policy. Unfortunately, his policy advice — a carbon tax, a uniform carbon tax, and nothing but a carbon tax — has fallen on deaf ears. I therefore extend Nordhaus' seminal DICE model, first in an *ad hoc* way to account for the inefficacy that comes with carbon price heterogeneity, and second in a more structural way by introducing desk-maximizing bureaucrats, again expressing this in a measure of inefficacy. I calibrate both extensions, the former to carbon prices observed around the world, the latter to emissions in the European Union and predictions by models used by the IPCC. The extended DICE model shows higher emissions and more warming than the original one, by 1°C maybe 2°C.

Several things can and should be improved about the research presented above. Although static approximations to DICE often give practically the same results as the dynamically optimized model, it would be good to check that for the current extensions.

More importantly, numerical models are only as good as their calibration. Besides the undue emphasis on first-best analyses, Nordhaus has another unfortunate legacy: an emphasis of *ex ante* analysis. Starting his research well before there was climate policy, Nordhaus had little choice. We now have 30 years of experience in attempting to reduce emissions, but the economics literature has made too little use of that data (this is now beginning to change, see Aichele and Felbermayr, 2012; Bel and Joseph, 2015; Almer and Winkler, 2017; Fowlie *et al.*, 2018; Andersson, 2019; Kube *et al.*, 2019; Maamoun, 2019, Xiang and Lawley, 2019; Metcalf and Stock, 2020; Runst and Thoniparaanita, 2020 and there are a few papers on corruption and emissions too (Arminen and Menegaki, 2019)). The calibration used above is simple, perhaps too simple, and may suffer from omitted variable bias. The simple calibration (and the papers just referred to) does point to potentially serious problems with the models used for climate policy analysis.

Finally, alternative models of selfish bureaucrats should be tried. The specification used here was chosen for analytical tractability. Partial civil servants are not the only policy imperfection. The impact of regulatory capture and lobbying on climate policy should be studied too (Stigler, 1971; Krueger, 1974).

None of this diminishes Nordhaus' contributions to the economics of climate change. He, I am sure, would like nothing better than the field to move beyond his work. We should, building on the foundations he so expertly laid.

Acknowledgment

I am grateful to Robert Mendelsohn for organizing this book in honor of William Nordhaus. Robert and an anonymous referee had excellent comments on a previous version of this paper.

References

Aichele, R and G Felbermayr (2012). Kyoto and the carbon footprint of nations. *Journal of Environmental Economics and Management*, 63(3), 336–354, http://www.sciencedirect.com/science/article/pii/S0095069611001422.

Almer, C and R Winkler (2017). Analyzing the effectiveness of international environmental policies: The case of the Kyoto Protocol. *Journal of Environmental Economics and Management*, 82, 125–151, http://www.sciencedirect.com/science/article/pii/S0095069616304296.

Andersson, JJ (2019). Carbon taxes and CO_2 emissions: Sweden as a case study. *American Economic Journal: Economic Policy*, 11(4), 1–30, http://www.aeaweb.org/articles?id=10.1257/pol.20170144.

Arminen H and A Menegaki (2019). Corruption, climate and the energy-environment-growth nexus. *Energy Economics*, 80, 621–634, doi: 10.1016/j.eneco.2019.02.009.

Barnett, AH (1980). The Pigouvian tax rule under monopoly. *The American Economic Review*, 70(5), 1037–1041, www.jstor.org/stable/1805784.

Barrage, L (2018). Be careful what you calibrate for: Social discounting in general equilibrium. *Journal of Public Economics*, 160, 33–49, http://www.sciencedirect.com/science/article/pii/S0047272718300380.

Barrage, L (2020). Optimal dynamic carbon taxes in a climate–economy model with distortionary fiscal policy. *The Review of Economic Studies*, 87(1), 1–39, https://doi.org/10.1093/restud/rdz055.

Barrett, S (1994). Self-enforcing international environmental agreements. *Oxford Economic Papers*, 46, 878–894.

Bator, FM (1958). The anatomy of market failure. *The Quarterly Journal of Economics*, 72(3), 351–379, https://doi.org/10.2307/1882231.

Baumol, WJ (1972). On taxation and the control of externalities. *The American Economic Review*, 62(3), 307–322, http://www.jstor.org/stable/1803378.

Baumol, WJ and WE Oates (1971). The use of standards and prices for protection of the environment. *The Swedish Journal of Economics*, 73(1), 42–54, www.jstor.org/stable/3439132.

Bel, G and S Joseph (2015). Emission abatement: Untangling the impacts of the EU ETS and the economic crisis. *Energy Economics*, 49, 531–539, http://www.sciencedirect.com/science/article/pii/S0140988315001061.

Böhringer, C, TF Rutherford and RSJ Tol (2009). The EU 20/20/2020 targets: An overview of the EMF22 assessment. *Energy Economics*, 31, S268–S273, http://www.sciencedirect.com/science/article/pii/S0140988309001935.

Buchanan, JM (1969). External diseconomies, corrective taxes, and market structure. *The American Economic Review*, 59(1), 174–177, www.jstor.org/stable/1811104.

Carraro, C and D Siniscalco (1992). The international dimension of environmental policy. *European Economic Review*, 36, 379–387.

Clarke, L, K Jiang, K Akimoto, MH Babiker, GJ Blanford, KA Fisher-Vanden, JC Hourcade, V Krey, E Kriegler, A Loeschel, DW McCollum, S Paltsev, S Rose, PR Shukla, M Tavoni, D van Vuuren and B Van Der Zwaan (2014). *Assessing Transformation Pathways*. Cambridge: Cambridge University Press.

Dales, JH (1968). Land, water, and ownership. *The Canadian Journal of Economics/Revue canadienne d'Economique*, 1(4), 791–804, www.jstor.org/stable/133706.

d'Arge, R (1979). Climate and economic activity. In *Proc. World Climate Conf.*, World Meteorological Organization, Geneva, pp. 652–681.

Fischer, C (2008). Emissions pricing, spillovers, and public investment in environmentally friendly technologies. *Energy Economics*, 30, 487–502.

Fowlie, M, M Greenstone and C Wolfram (2018). Do energy efficiency investments deliver? Evidence from the Weatherization Assistance Program. *The Quarterly Journal of Economics*, 133(3), 1597–1644, https://doi.org/10.1093/qje/qjy005.

Goulder, LH (1995). Environmental taxation and the double dividend: A reader's guide. *International Tax and Public Finance*, 2(2), 157–183.

Goulder, LH, IWH Parry, RC Williams Iii and D Burtraw (1999). The cost-effectiveness of alternative instruments for environmental protection in a second-best setting. *Journal of Public Economics*, 72(3), 329–360, doi: 10.1016/S0047-2727(98)00109-1.

Haites, E (2016). Experience with linking greenhouse gas emissions trading systems. *WIREs Energy and Environment*, 5(3), 246–260, http://doi.org/10.1002/wene.191.

Krueger, AO (1974). The political economy of the rent-seeking society. *The American Economic Review*, 64(3), 291–303, http://www.jstor.org/stable/1808883.

Krutilla, K (1991). Environmental regulation in an open economy. *Journal of Environmental Economics and Management*, 20(2), 127–142, doi: 10.1016/0095-0696(91)90046-L.

Kube, R, K von Graevenitz, A Löschel and P Massier (2019). Do voluntary environmental programs reduce emissions? EMAS in the German manufacturing sector. *Energy Economics*, 84, doi: 10.1016/j.eneco.2019.104558.

Lyon, RM (1990). Regulating bureaucratic polluters. *Public Finance Quarterly*, 18(2), 198–220, http://doi.org/10.1177/109114219001800204.

Maamoun, N (2019). The Kyoto Protocol: Empirical evidence of a hidden success. *Journal of Environmental Economics and Management*, 95, 227–256, http://www.sciencedirect.com/science/article/pii/S0095069618300391.

Metcalf, GE and JH Stock (2020). Measuring the macroeconomic impact of carbon taxes. *American Economic Review Papers and Proceedings*, 110, 101–106.

Montgomery, WD (1972). Markets in licences and efficient pollution control programs. *Journal of Economic Theory*, 5, 395–418.

Niskanen, W (1971). *Bureaucracy and Representative Government*. Chicago & New York: Aldine-Atherton.

Nordhaus, WD (1975). Can we control carbon dioxide? Report, International Institute for Applied Systems Analysis.

Nordhaus, WD (1977). Economic growth and climate: The case of carbon dioxide. *American Economic Review*, 67(1), 341–346.

Nordhaus, WD (1982). How fast should we graze the global commons? *American Economic Review*, 72(2), 242–246.

Nordhaus, WD (1991a). A sketch of the economics of the greenhouse effect. *American Economic Review*, 81(2), 146–150.

Nordhaus, WD (1991b). To slow or not to slow: The economics of the greenhouse effect. *Economic Journal*, 101(444), 920–937.

Nordhaus, WD (1992). An optimal transition path for controlling greenhouse gases. *Science*, 258, 1315–1319.

Nordhaus, WD (1993). Rolling the 'DICE': An optimal transition path for controlling greenhouse gases. *Resource and Energy Economics*, 15(1), 27–50.

Nordhaus, WD (1997). Discounting in economics and climate change — An editorial comment. *Climatic Change*, 37, 315–328.

Nordhaus, WD (2007). A review of the Stern Review on the economics of climate change. *Journal of Economic Literature*, 45(3), 686–702.

Nordhaus, WD and Z Yang (1996). RICE: A regional dynamic general equilibrium model of optimal climatechange policy. *American Economic Review*, 86(4), 741–765.

Oates, WE and DL Strassmann (1978). The use of effluent fees to regulate public sector sources of pollution — An application of the Niskanen model. *Journal of Environmental Economics and Management*, 5(3), 283–291, http://www.sciencedirect.com/science/article/pii/0095069678900141.

Oates, WE and DL Strassmann (1984). Effluent fees and market structure. *Journal of Public Economics*, 24(1), 29–46, http://www.sciencedirect.com/science/article/pii/0047272784900033.

Patuelli, R, P Nijkamp and E Pels (2005). Environmental tax reform and the double dividend: A meta-analytical performance assessment. *Ecological Economics*, 55(4), 564–583.

Pigou, A (1920). *The Economics of Welfare*. London: Macmillan.

Ranson, M and RN Stavins (2016). Linkage of greenhouse gas emissions trading systems: Learning from experience. *Climate Policy*, 16(3), 284–300, http://doi.org/10.1080/14693062.2014.997658.

Rehdanz, K and RSJ Tol (2005). Unilateral regulation of bilateral trade in greenhouse gas emission permits. *Ecological Economics*, 54, 397–416.

Ricke, K, L Drouet, K Caldeira and M Tavoni (2018). Country-level social cost of carbon. *Nature Climate Change*, 8(10), 895–900, http://doi.org/10.1038/s41558-018-0282-y.

Runst, P and A Thoniparaanita (2020). Dosis facit effectum — Why the size of the carbon tax matters: Evidence from the swedish residential sector. *Energy Economics*, 104898, http://www.sciencedirect.com/science/article/pii/S0140988320302383.

Stigler, GJ (1971). The theory of economic regulation. *The Bell Journal of Economics and Management Science*, 2(1), 3–21, http://www.jstor.org/stable/3003160.

Stiglitz, JE (2019). Addressing climate change through price and non-price interventions. *European Economic Review*, 119, 594–612, doi: 10.1016/j.euroecorev.2019.05.007.

Tol, RSJ (2018). The economic impacts of climate change. *Review of Environmental Economics and Policy*, 12(1), 4–25, http://doi.org/10.1093/reep/rex027.

Tol, RSJ (2019). A social cost of carbon for (almost) every country. *Energy Economics*, 83, 555–566, http://www.sciencedirect.com/science/article/pii/S014098831930218X.

van Heerden, JH, R Gerlagh, JN Blignaut, M Horridge, S Hess, R Mabugu and M Mabugu (2006). Searching for triple dividends in South Africa: Fighting CO_2 pollution and poverty while promoting growth. *Energy Journal*, 27(2), 113–141.

Weitzman, ML (1974). Prices vs. quantities. *The Review of Economic Studies*, 41(4), 477–491, https://doi.org/10.2307/2296698.

Xiang, D and C Lawley (2019). The impact of British Columbia's carbon tax on residential natural gas consumption. *Energy Economics*, 80, 206–218, http://www.sciencedirect.com/science/article/pii/S0140988318304869.

CHAPTER 7

CLIMATE CHANGE AND EXTERNALITY

ZILI YANG

Department of Economics, Binghamton University
State University of New York
4400 Vestal Parkway East, Binghamton, NY 13902, USA
zlyang@binghamton.edu

Climate change is an externality phenomenon. The DICE/RICE models are IAMs that treat climate change as an externality explicitly. Such a feature of DICE/RICE is recognized by the Nobel Committee and is one of the primary reasons for its influence. This paper argues the essentiality of incorporating external effects of climate change in our understanding of climate change from a socio-economic perspective; points out the biases of missing the externality elements in climate change economics; outlining the crucial role of externality in IAM modeling.

Keywords: Climate change; externality; integrated assessment model (IAM); the DICE and RICE models.

1. Introduction

In 2018, Professor William D. Nordhaus received the Nobel Prize in economics "for integrating climate change into long-run macroeconomic analysis." Summarizing the achievements of Nordhaus and Romer, the Nobel Committee points out that "both of them have emphasized externalities in their analysis of desirable long-run outcomes, thus pointing to a potentially important role for economic policy and offering new guidance for its design." (The Committee for the Prize in Economic Sciences in Memory of Alfred Nobel, 2018). Climate change is a complicated global externality phenomenon that affects economies and societies. Just like most other environmental issues, we should characterize and model climate change as an externality. Nordhaus' contributions to climate change economics have a clear thematic tone of externality.

Since the pioneering works by Nordhaus and other scholars in the late 1980s, climate change research has thrived from socio-economic perspectives. The economics of climate change has become a robust and active subfield of environmental economics. The appearance of a dedicated journal *Climate Change Economics* is one

This chapter was originally published in Climate Change Economics, Vol. 11, No. 4, December 2020, published by World Scientific Publishing, Singapore. Reprinted with permission.

such example. Among thousands of papers and books related to climate change's socio-economic aspects, some treat climate change as an externality; many do not. Scholars, primarily when they focus on some highly specific topical issues, often place the externality characteristics of climate change into the backburner. We would argue that such an approach might be biased and unappealing scholarly under certain circumstances. Modeling climate change from an economic angle as an externality phenomenon is the balanced and correct perspective.

Externalities are ubiquitous in our socio-economic lives. There are many definitions of externalities. One of them is the "indirect effect of a consumption activity or a production activity on the consumption set of a consumer, the utility function of a consumer or the production function of a producer" given by Laffont (2008). All pollution, including climate change, fit this definition. Regional economic activities generate GHG emissions separately. The emissions are flows of externality. The concentration of GHG emissions causes the temperature to increase. The temperature increase is the stock of externality. As a result, global warming affects all regions' welfares. Externality implies a market failure. Climate change is an externality phenomenon. Thus, climate change has all signatures associated with market failure from an economic perspective. The presence of externality leads to many predicted and unexpected complications in climate change research.

In this paper of commemorating Nordhaus' Nobel Prize winning achievement, we would explore the relationship between climate change and its externality characteristics, and highlight the significance of externality in climate change research. The remaining parts of the paper are as follows. The following section contains the arguments that the DICE and RICE models are externality models; thus, the externality characterization of climate change is one of the core ideas of Nordhaus' works. In Sec. 3, the deficiencies of not modeling climate change as an externality are analyzed. Section 4 discusses the necessity of incorporating externality in the comprehensive approach to climate change, particularly in IAMs. Section 5 sums up our main points.

2. The DICE and RICE Models as Externality Models

As pointed out by the Nobel Committee, Nordhaus' methodological contributions to climate change research are the development of the DICE and RICE models. The DICE model (Nordhaus, 1994) and the RICE model (Nordhaus and Yang, 1996) are two influential integrated assessment model (IAM) for climate change. Since their initial inception in the mid-1990s, the models have experienced several rounds of updating and revisions. "They still remain the workhorse models for climate economics all over the world." (The Committee for the Prize in Economic Sciences in Memory of Alfred Nobel, 2018). There are many reasons for the models' influence and popularity One fundamental reason is that they capture one of climate economics' key features in concise modeling language and logic. This key feature is the externality.

The DICE model is a Solow optimal growth model of the global economy with CO_2 emissions being a by-product of economic growth.[1] This by-product has negative impacts on economic growth. From modeling perspective, DICE is an optimal control model. The optimal time paths of control (decision) variables, and the paths of state variables, are identified numerically, to solve the model. In DICE, the control variables are the capital investment rate $I(t)$ (just as in a classical Solow model) and CO_2 control rate $\mu(t)$. The optimal path of $\mu(t)$ determines the percentage of emission mitigation, balancing the mitigation costs and climate damage. On the surface, the DICE model's optimal solution is merely a result of cost-benefit tradeoffs, and the external effects of climate change are not present. Still, we can argue that climate change is fully internalized in DICE's optimal solutions. Even in this setup, climate change is an "intertemporal" or "intergenerational" externality in the DICE model. CO_2 emissions by the current generation affect the welfare of future generations (without their consent) through the accumulation of CO_2 concentration in the atmosphere. This process satisfies the definition of (intergenerational) externality. The perpetual "decision-maker" in the DICE model internalizes this one-directional externality over many generations and comes up with efficient $\mu(t)$ for all generations.

The RICE model is a multi-region counterpart of the DICE model. This regional disaggregation of the DICE model reveals the full-fledged feature of climate externality. In the RICE model framework, climate externality is the main theme connecting the components of the model. Regional economic growth generates CO_2 emissions $E(n, t)$ in the RICE model's economic module. Then, the aggregate emission causes temperature changes in the RICE model's carbon cycle module, or $\sum_n E(n, t) \rightarrow \Delta T(t)$. Temperature changes enter every region's utility function for its negative economic impacts, or $\Delta T(t)$ is a common argument in every $U(n)$. Region i's economic activities affect region j's welfare through external effect $E(i, t), \forall (i, j)$. This feature is precisely externality.

The object function of the RICE model is a weighted social welfare function that sums up all regions' utility functions. The optimal solutions of the RICE model fully internalize the climate externality. Different social welfare weights map the optimal solutions to the entire efficiency frontier of the model. The "social planner" of the model executes the decision based on all regions' full cooperation. Like other situations involving negative externalities, the free market does not provide externality efficiently, and the public "bad" is always over-supplied. In climate change, regional CO_2 emissions are excessive without policy interventions. Such outcomes are typical characteristics of market failure caused by externality. To have a complete "picture" of climate externality, we should know the model's noncooperative Cournot–Nash equilibrium or the outcome of market failure. The RICE model's Cournot–Nash equilibrium is identified, and the Nobel Committee recognizes this finding (The

---

[1]The mathematical structures of the DICE and RICE models are available in the literature and not described here.

Committee for the Prize in Economic Sciences in Memory of Alfred Nobel, 2018).[2] In the subsequent development of the RICE model, the solution concepts related to externality are richer. Particularly, the Lindahl equilibrium of the RICE model has been identified (Yang, 2020). In an economy with externalities, the Lindahl equilibrium is as important as the Walrasian equilibrium to an Arrow–Debreu economy. Therefore, despite its complicated structure, the RICE model is a "textbook" like model of externality provision with all important solution concepts identified.

Modeling climate change as an externality in RICE reflects a balanced and perceptive understanding of climate change by economists. None of the economic models can be omnipotent and resolve all concerned problems. A good model captures the essential characteristics of the issues. For climate change, the essential characteristics is the externality. The modeling approach of DICE/RICE is just right. The model incorporates CO_2 emissions (the generation of externality) and climate damage (the impact of externality) into a single framework. Such a complete "loop" forms a well-defined problem of externality provision. Balancing mitigation costs and mitigated climate damage composes a rigorous cost-benefit analysis (CBA) problem involving external effects, and the DICE/RICE model handle this CBA elegantly. Finally, the DICE/RICE model combines the external effects of regional CO_2 emissions and public "bad" features of temperature increase in the CBA. The demarcation of the efficient solutions of a social planner's problem and the inefficient noncooperative Cournot–Nash equilibrium in the RICE model is an exemplar research framework of externality provision.

3. Biases from the Absence of Externality

The DICE/RICE models represent a distinctive approach in climate change economics, and there are several major research groups and scholars have adopted RICE modeling philosophy. These models combine GHG emissions and climate impacts in an integrated framework and GHG emissions generates external effects of across regions in these models. On the other hand, many IAM models do not adopt this approach, as we pointed out at the beginning. The IAMs constructed with simple simulations or computable general equilibrium (CGE) methods usually do not close the "loop" connecting GHG emissions and climate impacts in an integrated framework.

IAMs are utilized to assess global and regional GHG mitigation costs. In an IAM based on CGE, the exercise is conducted in the following way. A mitigation target is treated as a quantity (or price) constraint imposed on the system. The economy adjusts to meet such a target accordingly, and a new equilibrium met the mitigation restriction is found. This new equilibrium represents a "cost-effective" mitigation outcome and its

---

[2]The author of this paper still remembered vividly that during the early development of RICE, Nordhaus pointed out the necessity to find the Nash equilibrium to me. We discovered the algorithm to identify the noncooperative Cournot–Nash equilibrium, and the method is still used today for identifying the noncooperative solutions of externality provisions numerically.

welfare impacts are assessed. Because CGE models are calibrated with a social accounting matrix (SAM), climate damage is tough to be built in the model directly. Whether a cost-effective policy is an optimal policy or not is a question because the mitigated impacts are not considered in the mitigation. Furthermore, the impacts on a region are caused by the emissions from all regions in the world. If this damage link of the "loop" is not connected, the calculation of mitigation costs at regional or sectorial levels is biased. When a region is vulnerable to adverse climate impacts, the opportunity cost of mitigation is under-estimated by CGE. They may want to mitigate more GHG emissions (partially depends on other regions' mitigation efforts), and the real mitigation costs are lower than a CGE model suggested. The cases of over-estimation occur too.

Another problem of disregarding externality in climate change economics is the misconceptions regarding international cooperation. Cooperation or noncooperation in GHG mitigation is a strategic consideration of a sovereign country facing climate externality. Such consideration is mainly based on the cost-benefit of its mitigation efforts and external effects from other countries' mitigation decisions. Without specifying externality, we cannot characterize the nature of international cooperation correctly. In the literature, the term "cooperation" sometimes is just semantic. We do not ask what "cooperation" really means, and cooperation seems easy in some IAMs. A scenario simulation from a CGE can be labeled as a "cooperation" result or something else. How would such cooperation internalize climate externality and the free-riding disincentives to such cooperation are not clear. Ignoring external effects renders an unrealistic optimistic view toward international cooperation. As we learned from the international negotiation of climate change responses in the past three decades, the reality has told a different story.

Third, treating climate change as an externality phenomenon offers two distinctive benchmark scenarios: the inefficient noncooperative Cournot–Nash equilibrium and the set of efficient outcomes (the weighted social optimums). Other policy-related scenarios can be developed from these two benchmarks. However, the benchmarks of IAMs based on CGE are blurred without such externality specifications. The benchmark of noncooperative Cournot–Nash equilibrium is absent in CGE models.[3] A simple term "baseline" may imply different scenarios in different IAM models. Some baselines are "optimal" or equilibrium outcome without any GHG mitigations; some are the outcomes that fit certain *ad hoc* assumptions. In any case, the fundamental assumption of a CGE model is that the underlining economy is an Arrow–Debreu economy without externality. The "optimal" responses in a CGE model are "off mark" if ignoring GHG emissions' external effects.

The biases mentioned above are from missing the consideration of external or "spill-over" effects of climate externality. While the magnitude of such external effects

---

[3]The benchmark calibration of a CGE model is a market equilibrium (i.e., the SAM table) without externality. It is different from the noncooperative Cournot–Nash equilibrium in the presence of externality.

to a country and the world is subject to investigation, ignoring their presence is not correct. In econometrical applications, scholars seldom accept a biased estimation; in structural modeling of climate change, we should not take the misspecifications of climate change, i.e., omitting externality, for granted either.

4. The Role of Externality in IAMs for Climate Change

The critiques in Sec. 3 should not be interpreted as an indiscriminate denouncement of a large portion of climate research. Many specific research topics in climate change economics can put "externality" in the backburner, based on economists' "ceteris paribus." For example, the research topics such as "the impacts of temperature increase on the agriculture in country A" and "carbon mitigation policy and development of clean energy in region B" do not directly connect with the externality concept. Nevertheless, we would argue that ignoring externality in a comprehensive climate change framework is likely a mistake or nontrivial elapse.

Here the comprehensive framework refers to IAMs for climate change. IAM has been the mainstream research methodology in climate change economics. IAMs fulfill many research tasks in climate change research, from assessing GHG mitigation costs to evaluating various future impacts. Following the pioneering DICE/RICE endeavors, many IAMs have been built thereafter. Kelly and Kolstad (1999) did an excellent survey of the early works in IAMs. Yang *et al.* (2015) provide a broad overview of the *status quo* of IAMs in a bibliographic format. Of the IAMs covered by the above two sources, only a small portion treats climate change as an externality phenomenon explicitly. We may call this minority of IAMs as the "RICE-type" models. The majority of IAMs are based on the CGE methodologies.

The methodological divide between the "RICE-type" and CGE models is wide. It is the notion of externality that separates the two. The IAMs based on the CGE approach can fulfill many useful research tasks. They are the leading forces in various scenario analyses initiated by IPCC, such as IPCC (2000, 2018); they respond promptly to the policy mandates by UNFCCC with detailed assessments; they contribute extensively to the literature of climate change economics. While it is not fair to criticize the CGE methodologies here as they are continuing controversies in the literature, we would like to point out that the problems related to the absence of externality as briefed in Sec. 3 are present in many IAMs using the CGE approach. Therefore, the reservations to CGE models should be in place.

The "RICE-type" IAM models have caveats and shortcomings too. The reasonable critiques of DICE/RICE have been in the literature since the models' early inception. A common criticism focuses on the simplicity of DICE/RICE. Other "RICE-type" models overcome some shortcomings of RICE by adding valuable structures and mechanisms, such as the WITCH model (Bosetti *et al.*, 2006). Nonetheless, the plus side of simplicity in modeling is transparency and portability of the model. The popularity of DICE/RICE is partially due to its transparency and portability. In all,

DICE/RICE captures the essential characteristics of climate change in a simple framework: it is climate externality. The simplicity of the RICE model highlights and emphasizes the externality nature of climate change.

Characterizing climate change as an externality phenomenon is the correct way to build an IAM. Having justified the essentiality of including externality in IAMs, we would also like to point out that treating climate change as an externality phenomenon opens the door to many interesting economic research topics. The game-theoretic modeling of climate change is one of them. To explore the complicated mechanical nesting structure of a CGE model or to capture some behavioral patterns of economic agents are choices made by scholars. Neither is perfect, but we should consider what are foresaken. The "RICE-type" models allow the game-theoretic approach and capture specific "behavioral" patterns of regions. The noncooperative Cournot–Nash equilibrium solution in the original RICE model is an example.

Cheating and free-riding by polluters are unavoidable problems in the implementation of environmental policies, climate change is no exception. The best approach to study such a problem is through game-theoretic modeling. In a global environmental problem like climate change, overcoming cheating and free-riding is through international environmental agreements (IEAs). Cooperation is the means to internalize climate externality fully. To forge global cooperation in response to climate change, the regions crucially rely on self-interesting incentives. The external effect of other regions' GHG mitigation strategies is a critical factor in assessing such incentives. Cooperative game theory or coalition theory is the research methodology to deal with such situations. The "RICE-type" models allow coalitional solutions and answer some questions raised from the game-theoretic angle. For example, all participating models in a comparative study of climate coalitions are "RICE-type" models that treat climate change as an externality (Lessmann *et al.*, 2015). The "RICE-type" models show the capability of handling the strategic aspect of climate change.

5. Concluding Remarks

Long-run macroeconomic analysis at the country and global level should consider the potential long-term impact of climate change. Climate change is an externality phenomenon, and it might be that "climate change is the mother of all externalities" (Tol, 2009). Therefore, scholars in climate change economics have paid particular attention to externality characteristics of climate change. The DICE/RICE modeling framework pioneered this approach.

The DICE/RICE model's unique advantage is that it includes all major components of climate change economics into a unified and straightforward modeling framework. For DICE, it is a dynamic CBA at a global level for an aggregate intergenerational externality problem; for RICE, it treats climate change as a global *externality* and solves for various solutions (efficient social optimums versus inefficient Cournot–Nash equilibrium, "the first-best" versus "the second-best" scenarios). The DICE/RICE model is the first IAM that closes the "loop" for CBA in the presence of climate externality. In contrast, the majority of IAMs based on CGE does not have damage built in directly. The subsequent IAMs that treat climate change as an externality phenomenon and incorporate mitigation cost and climate change into a single framework are "RICE-type" models. The DICE/RICE model's transparency and simplicity emphasize the role of externality in climate change.

Many research topics in environmental economics and climate change economics are involved with the externality concept. As an economist who has learned extensively from the DICE/RICE model, the author of this paper has been attracted to the externality feature of climate change and other environmental problems (Yang, 2020). There are many unresolved issues related to climate change and externality in front of us. As the Nobel Committee pointed out, it plays a vital role in economic policy. We study climate change from an economic perspective as it is — externality.

Acknowledgments

The author thanks the financial support of the National Key R&D Program of China (No. 2016YFA0602603) and the Whitney and Betty MacMillan Center for International and Area Studies at Yale.

References

Bosetti, V, C Carraro, M Galeotti, E Massetti and M Tavoniet (2006). WITCH: A world induced technical change hybrid model. *The Energy Journal*, 27(1), 13–37.

IPCC (2000). Emissions scenarios. Cambridge, UK: Cambridge University Press.

IPCC (2018). Global warming of 1.5°C. Cambridge, UK: Cambridge University Press.

Kelly, DL and CD Kolstad (1999). Integrated assessment models for climate change control. In *International Yearbook of Environmental and Resource Economics 1999/2000: A Survey of Current Issues*, H Folmer and T Tietenberg (eds.), pp. 171–197. Cheltenham, UK: Edward Elgar.

Laffont, JJ (2008). Externalities. In *The New Palgrave: A Dictionary of Economics*, Vol. 3, pp. 192–194, 2nd edn., SN Durlauf and LE Blume (eds.). The Macmillan Press.

Lessmann, K, U Kornek, V Bosetti, R Dellink, J Emmerling, J Eyckmans, M Nagashima, H Weikard and Z Yang (2015). The stability and effectiveness of climate coalitions: A comparative analysis of multiple integrated assessment models. *Environmental and Resource Economics*, 62, 811–834.

Nordhaus, WD (1994). *Managing the Global Commons: The Economics of Climate Change*. Cambridge, MA: MIT Press.

Nordhaus, WD and Z Yang (1996). A regional dynamic general-equilibrium model of alternative climate-change strategies. *American Economic Review*, 86(4), 741–765.

The Committee for the Prize in Economic Sciences in Memory of Alfred Nobel (2018). Economic growth, technological change, and climate change. Available at https://www. nobelprize.org/uploads/2018/10/advanced-economicsciencesprize2018.pdf.

Tol, R (2009). The economic effects of climate change. *Journal of Economic Perspectives*, 23(2), 29–51.

Yang, Z, Y Wei and Z Mi (2015). Integrated assessment models for climate change. In *Oxford Bibliographies in Environmental Science*, E Wohl (ed.). New York: Oxford University Press, http://www.oxfordbibliographies.com/view/document/obo-9780199363445/obo-9780199363445-0043.xml.

Yang, Z (2020). *The Environment and Externality: Theory, Algorithms, and Applications.* Cambridge, UK: Cambridge University Press.

© 2021 World Scientific Publishing Company
https://doi.org/10.1142/9789811247699_008

CHAPTER 8

CLIMATE CLUBS WITH TAX REVENUE RECYCLING, TARIFFS, AND TRANSFERS

DAIGEE SHAW* and YU-HSUAN FU

Institute of Economics, Academia Sinica, Taiwan
*dshaw@sinica.edu.tw

The E3ME-FTT model is applied to assess the impacts of alternative climate club structures. We consider two kinds of climate club memberships: the World Climate Club (WCC), where every country in the world joins the club, and the Core Climate Club (CCC), with seven likely club members: EU + 5, Japan, South Korea, Canada, Brazil, Mexico, and Australia. First, we find that both the WCC and domestic revenue-neutral recycling matter a lot. The global CO_2 emissions in 2050 could be reduced by 50% from BAU under the WCC. With domestic revenue-neutral recycling, there will be large positive impacts on GDP under both the WCC and the CCC. Secondly, the negative effects of trade sanctions on cumulative global GDP and global CO_2 emissions make it unwelcome to be used as part of the club design. Lastly, the introduction of international transfers will result in a win–win solution that will not only increase the cumulative global GDP and reduce global CO_2 emissions but also enhance the equality among club members and induce more likely participation in the climate club.

Keywords: Climate club; carbon tax; revenue recycling; tariff; international transfer.

1. Introduction

Climate change is an existential and urgent threat that the whole of humanity is facing. To reduce this existential threat, the Paris Agreement of 2016 set up a long-term global average temperature target of 2°C above pre-industrial levels intending to pursue a further target of 1.5°C.[1]

Although climate change is an urgent existential threat to the whole of humanity, its very nature as a global and intergenerational public externality makes individuals and countries reluctant to join international climate agreements, such as the United Nations Framework Convention on Climate Change (UNFCCC), in order to make the significant contributions necessary to mitigate the threat. To facilitate international

---

This chapter was originally published in Climate Change Economics, Vol. 11, No. 4, December 2020, published by World Scientific Publishing, Singapore. Reprinted with permission.

[1]Nordhaus (1975, 1977) was the first to suggest using 2°C as a critical limit for climate policy in order to keep the climate within the normal range of long-term climatic variation. This limited range of variation is important due to the fact that human society and culture have developed within or below this climate range over the last several hundred thousand years.

cooperation on climate change, minilateral solutions to climate change, such as climate clubs, have been proposed by Barrett (2003) and Victor (2006, 2011) and studied extensively in recent years. Since Nordhaus (2015) analyzed the benefits of different climate club configurations, research on the effectiveness of climate clubs has been booming. The settings of climate clubs that Nordhaus (2015) has tested using the Coalition-DICE model include four target carbon taxes for club members and 11 tariff rates for imports from nonmembers. Nordhaus (2015) found that all configurations can achieve net gains in welfare for all major emitters, in addition to emission reductions, for all major emitters.

Several studies focus on which countries may be the initiators of climate clubs and the importance of some countries. Hovi *et al.* (2019) used agent-based simulations to examine the effectiveness of climate clubs and found that a climate club is likely to persist and grow if initiated by the USA and the European Union. Sprinz *et al.* (2018) analyzed the impact that the lack of USA participation may have on membership and the emissions. However, Camuzeaux *et al.* (2020) emphasized the importance of India because India is poised to take China's position by around 2030 even though the USA and China are now the two largest emitters of greenhouse gases.

The decision of a country to join or not to join a climate club will also affect the macroeconomy across countries. However, there are not many empirical studies on the macroeconomic assessment of climate clubs across countries. One exception is Paroussos *et al.* (2019), who provided a quantitative macroeconomic evaluation of the costs and benefits that would be associated with climate clubs. They structured the club with two more benefits for developing countries: enhanced technological diffusion and the provision of low-cost climate finance.

Several studies have also analyzed whether a uniform percentage tariff on all imports from nonparticipating countries is an effective mechanism to induce participation. Böhringer *et al.* (2016) used a Nash equilibrium game model to analyze the strategic value of carbon tariffs and stated that tariffs induce the cooperation of nonparticipants. Nordhaus (2015) examined the effect of using a uniform tariff mechanism for different carbon price levels.

The existing literature has already assessed several aspects of climate clubs, but there are still several gaps that need to be filled. First, the possible positive effect of carbon tax revenue recycling on the economy has not been considered in the literature on climate clubs. Second, the innovation and progress of mitigation technologies are crucial for this very long-term problem of climate change. Third, to induce participation from low-emission developing countries, equalities and inequalities among countries in both emissions and income are critical issues because it is generally agreed that those countries whose emissions are higher should take more responsibility for climate change. Since it is difficult for low-income countries to invest in and develop low-carbon technology, financial transfers are necessary for facilitating climate clubs. Fourthly, nonmembers that are penalized by members with high tariffs will not accept the penalty quietly, and may divert trade from club members to other nonmembers or trigger trade wars. However, few papers have studied these responses.

The purpose of our study is to address these gaps using the E3ME-FTT model. We consider climate club configurations whereby members charge a uniform carbon tax and recycle all of the tax revenue, and where members charge a uniform percentage tariff on all imports from nonparticipating countries and allow financial transfers among higher and lower emission countries. The E3ME-FTT model can model the innovation and progress of mitigation technologies in the long run and enable countries to divert trade among trading partners facing different tariffs. The remainder of this paper is organized as follows: Section 2 introduces the E3ME-FTT model that we used. Section 3 presents a brief description of our design of climate clubs. Section 4 describes the scenarios. Section 5 provides simulation results. Section 6 concludes the paper.

2. Methodology

The E3ME model (energy–environment–economy global macro-economic model) is a global, macro-econometric model designed to address significant economic and environmental policy challenges and has been developed by Cambridge Econometrics over the last 25 years. Cambridge Econometrics was founded in 1978 as a commercial spin-off from the University of Cambridge to take forward the pioneering work of Professor Sir Richard Stone. The primary research field of Cambridge Econometrics is the applied economic analysis of energy, environmental, and economic policies. The first version of the E3ME model was built by an international European team under a succession of European Commission research projects (which were completed in 1999) and is now widely used for policy assessment. The E3ME model covers 59 regions' economic and energy systems and the environment, with 43 economic sectors in each region (69 sectors in Europe), and has the capability to project forward annually up to 2050.

There are three essential features of the E3ME model. First, it covers detailed regions and sectors, enabling a thorough analysis of sectoral and country-level effects from a wide range of scenarios. Second, E3MEs econometric approach is based on the macro-econometric model and provides a robust empirical basis for analysis. The E3ME model can fully assess both short- and long-term impacts and is not limited by many of the restrictive assumptions common to Computable General Equilibrium (CGE) models. For instance, rebound effects are considered in the E3ME model. Finally, the E3ME model contains an integrated treatment of the world's economies, energy systems, emissions, and material demands. This enables it to capture two-way linkages and feedbacks between these components.

Furthermore, to link technology development and macroeconomic performance, there is a model of Future Technology Transformations (FTT) in E3ME. FTT models technology transformation from an engineering background, using a novel framework for the dynamic selection and diffusion of innovations, initially developed by Mercure (2012). There are several FTT models, including the power sector, road transportation, household heating, steel, etc. The approach is based on the Lotka–Volterra equations, also known as the predator–prey equations, which are used to describe the dynamics of biological systems in which two species interact, one as a predator and the other as prey. FTT sub-models have applied the concept to project future technology transformations. FTT sub-models are based on a decision-making core for investors who must choose from a list of available technologies. 'Levelized' cost distributions (including capital and running costs) are fed into a set of pairwise comparisons, which are conceptually similar to a binary logit model. In the end, FTT sub-models will feedback these results to the E3ME macro-econometric model and calculate the impacts on the economy. Therefore, the E3ME-FTT integrated assessment model is able to simulate the future technology transformations based on both top-down economic analyses and bottom-up engineering analyses, giving rise to highly accurate empirical results. By contrast, CGE models and other macro-econometric models are unable to model technology using a bottom-up approach.

Cambridge Econometrics applied the E3ME model to analyze the impacts of EU climate policies, for instance, "Macroeconomic analysis of the employment impacts of future EU climate policies" (Pollitt *et al.*, 2015) and "EU climate and energy policy beyond 2020: Is a single target for GHG reduction sufficient?" (Smith *et al.*, 2019). As for Asia, there are research teams in Japan and Korea. Most of their research has been published in the following two books: *Low-carbon, Sustainable Future in East Asia: Improving energy systems, taxation and policy cooperation* (Lee *et al.*, 2015) and *Energy, Environmental and Economic Sustainability in East Asia: Policies and institutional reforms* (Lee *et al.*, 2019).

Many empirical studies on environmental economics have applied computable general equilibrium (CGE) models, which are based on neoclassical economic theory. The critical assumption is perfect knowledge and that people decide rationally. CGE models also assume full employment. The model results are long-term impacts and usually do not allow for short-term dynamic (or transition) outcomes. By contrast, the E3ME model takes limited knowledge into consideration, which means the decision will be affected by past choices.

There are two reasons why we use the E3ME model. The first relates to path dependence or the anchoring effect, which means that the decision we make today is profoundly affected by past choices, even though the circumstances in the past are currently irrelevant. The E3ME model is based on historical data and simulates the future. Therefore, it is more realistic and is able to provide empirical outcomes. Second, the E3ME model can estimate the impacts of aggregate demand due to technological progress. Since a CGE model is based on neoclassical economics theory, it is difficult for it to capture aggregate demand shocks from low-carbon technological innovation. The reason for this is that the demand-side effects are short-term impacts under neoclassical economics theory, while the CGE model's outcome is long-term equilibrium. By contrast, E3MEs origins lie in post-Keynesian economics theory and it is a demand-driven model. In the E3ME model, implementing deep-decarbonization policies increases the investment in the power sector that, in turn, increases aggregate demand to achieve economic growth.

3. The Design of Climate Clubs

Based on the literature review above, we find that the proper design of climate clubs should include four parts: a uniform carbon tax, domestic revenue-neutral recycling, trade sanctions, and transfers. First, if one is to achieve the net-zero CO_2 emissions by 2050, a higher carbon tax is a must. In order to reform the UK's approach to carbon pricing to make the net-zero target, Burke *et al.* (2019) applied the Hotelling rule to calculate the shadow price of carbon, which rises at an interest rate of 3.8%, to reach the price of the backstop technology by 2050. Negative emissions technologies are seen as backstop technologies of complete decarbonization. We follow Burke *et al.* (2019) and use their shadow prices of carbon as the tax rate of a uniform tax on carbon emissions in our design of climate clubs. This suggests that the carbon tax rate is €63/tCO_2 in 2020, reaching €91/tCO_2 in 2030 and €192/tCO_2 in 2050.

Second, as for domestic carbon tax revenue recycling, there are several available approaches built into the E3ME model, including reducing VAT, enterprise and household income tax rates, and the social security payment accruing to employers; increasing social security and welfare payments to low-income households; and uniform lump-sum transfers. Of these, we use lump-sum transfers because this approach is simple and can meet the goals of income distribution equality and economic growth. Furthermore, several studies have proposed lump-sum transfers as an ideal way of revenue recycling (Baker III *et al.*, 2017; Klenert *et al.*, 2018; Chewpreecha and Lee, 2015).

Third, in the case of trade sanctions, we follow Nordhaus (2015). Nordhaus (2015) found that there were no stable cooperative emission reduction coalitions without appropriate trade sanctions against nonparticipants. He thus proposed using a uniform percentage tariff to increase participation in climate clubs. The participation rate rises with the tariff rate. In our study, we use a 10% tariff.

Fourth, we consider using a portion, say, 10%, of the carbon tax revenue for international transfers among club members. Based on the principles of "One Human — One Emission Right" (Ekardt and von Hovel, 2009) and "Emissions Egalitarianism" (Torpman, 2019), below-average countries (recipients) will receive transfers from the above-average members (benefactors). We first calculate the CO_2 emissions per capita of each country and the club. The latter one is called the club average. Then we calculate 10% of the total carbon tax revenue of those benefactor countries and allocate the transfer responsibility to each benefactor depending on the differences between the country's CO_2 emissions per capita and the club average and its population size. The same approach is used to allocate the transfers each recipient will receive (see

Appendix A for the formulas used to calculate the transfers). For benefactors, their available domestic carbon tax revenues for recycling will be reduced by the amounts of transfers paid. For recipients, the transfers received will be used for government investments.

The design of transfers will help induce the participation of those low-income and low-emission countries and trigger emission reduction and innovation. To reduce transaction costs, the amounts of transfers are assessed in 2020, 2030, and 2040, and kept the same for every 10 years.[2]

4. Scenarios

First, we identify the members of a climate club. We consider two kinds of climate club memberships. The first one is referred to as the World Climate Club (WCC), where every country in the world joins the club. As for the second one, the Core Climate Club (CCC), we follow Martin and van den Bergh (2019), who developed a method to identify a group of seven 'likely' club members. The method utilizes four complementary criteria: (1) carbon independence, (2) public opinion regarding climate change, (3) government policy position, and (4) climate coalition membership to predict the likelihood-of-involvement from 15 countries[3] with the highest carbon emission rates. These seven likely club members are EU + 5 (the UK, Norway, Switzerland, Iceland, and Macedonia), Japan, South Korea, Canada, Brazil, Mexico, and Australia.

Second, we define nine scenarios for the two kinds of climate clubs. Table 1 presents the definitions of all of the nine scenarios, including the business-as-usual (BAU) scenario, two WCC scenarios, and six CCC scenarios. In the WCC, every country joins the climate club, implementing a uniform carbon tax from 2020 to 2050. In the WCC + R scenario, all of the carbon tax revenue is recycled through lump-sum

Table 1. Description of scenarios.

Scenarios	Description
Business-as-usual (BAU)	• Assuming no change in the regulatory program under consideration.
World Climate Club (WCC)	• Every country joins the WCC, implementing a uniform carbon tax. • No carbon revenue recycling.

---

[2]We use the population in 2020 to calculate the CO_2 emissions per capita of each country and of the club in 2020, 2030 and 2040, and the transfers, because a fixed basis of population would keep countries from using population increase as a strategic behavior (Ekardt and von Hovel, 2009).

[3]The term 'country' is used here to indicate all potential country members and the European Union (EU).

Table 1. (*Continued*)

Scenarios	Description
WCC + R	• Every country joins the WCC, implementing a uniform carbon tax. • 100% of the carbon tax revenue is recycled through lump-sum transfers.
Core Climate Club (CCC)	• Seven countries form the CCC, implementing a uniform carbon tax. • No revenue recycling.
CCC + R	• Seven countries form the CCC, implementing a uniform carbon tax. • 100% of the tax revenue is recycled through lump-sum transfers.
CCC + R + 10% tariffs	• Seven countries form the CCC, implementing a uniform carbon tax. • 100% of the tax revenue is recycled through lump-sum transfers. • 10% tariffs on imports from nonclub members.
CCC + R + 0.1 Transfer	• Seven countries form the CCC, implementing a uniform carbon tax. • For countries whose CO_2 emissions per capita is lower than the club average, 100% of the tax revenue is recycled through lump-sum transfers, and receive international transfers. • For countries whose CO_2 emissions per capita is higher than the club average, their available domestic carbon tax revenues for recycling will be reduced by the amounts of international transfers paid.
CCC + USA + R	• Eight countries (Core + USA) form the CCC, implementing a uniform carbon tax. • 100% of the tax revenue is recycled through lump-sum transfers.
CCC + Taiwan + R	• Eight countries (Core + Taiwan) form the CCC, implementing a uniform carbon tax. • 100% of the tax revenue is recycled through lump-sum transfers.

transfers to their citizens. No trade sanction is needed for the WCC and WCC + R scenarios. As for the CCC scenario, we consider alternative policy scenarios that have such additional features as a uniform carbon tax with or without tax revenue recycling, a uniform tariff, and transfers among high and low per-capita-emission members.

5. Modeling Results

We summarize the simulation results in Table 2 and Figs. 1–3. Table 2 shows the tradeoffs between the global environmental and macroeconomic impacts of different scenarios, while Fig. 1 shows the macroeconomic impacts for different countries under different scenarios of the CCC. Figure 2 shows the resources transferred among club members under the scenario of CCC + R + 0.1 Transfer in 2020, 2030, and 2040.

In Table 2, we only present those scenarios' global environmental impacts in terms of the global CO_2 emissions in 2050 and macroeconomic performances in terms of the global cumulative GDP (the change in GDP from BAU), among many other indicators assessed by the E3ME model, because the decision to join or not to join the club is a

Table 2. Modeling results of emissions and GDP.

Scenarios	Global CO_2 emissions in 2050 (billion tCO_2)	Global (discounted[1]) cumulative GDP (change from BAU (%))
BAU	44.3	—
World Climate Club (WCC)	23.3	−0.62%[2]
World Climate Club (WCC) + R	23.7	3.39%[2]
Core Climate Club (CCC)	42.7	−0.37%
CCC + R	41.1	0.29%
CCC + R + 10% tariffs	42.8	0.26%
CCC + R + 0.1 Transfer	40.8	0.30%
CCC + USA + R	38.6	0.67%
CCC + Taiwan + R	40.5	0.30%

Notes: [1]The discount rate used is 3%. [2]Due to the boundaries of the E3ME model, in the WCC and WCC + R scenarios, the model can only project up to 2038. Therefore, we use the same trend from 2020 to 2038 to estimate global CO_2 emissions in 2050. The cumulative GDP is cumulated from 2020 to 2038. All the other scenarios are from 2020 to 2045.

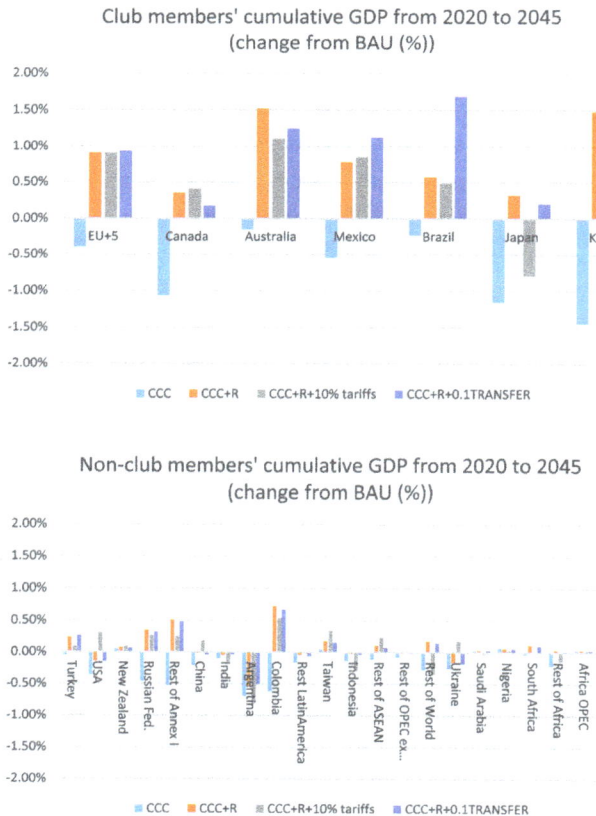

Figure 1. CCC members' and nonmembers' cumulative GDP from 2020 to 2045 (change from BAU (%)).

TRANSFER per year as a percentage of GDP (%)

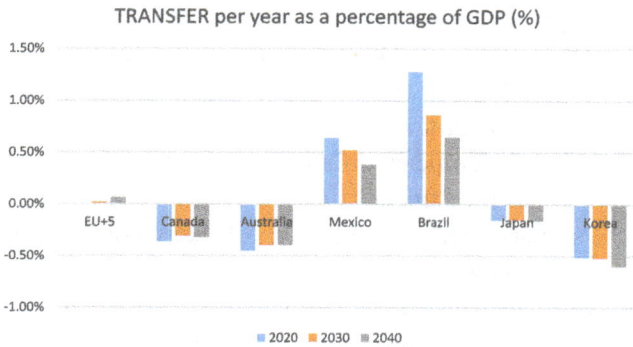

Figure 2. CCC members' transfers per year as a percentage of GDP in 2020, 2030, and 2040 (%).

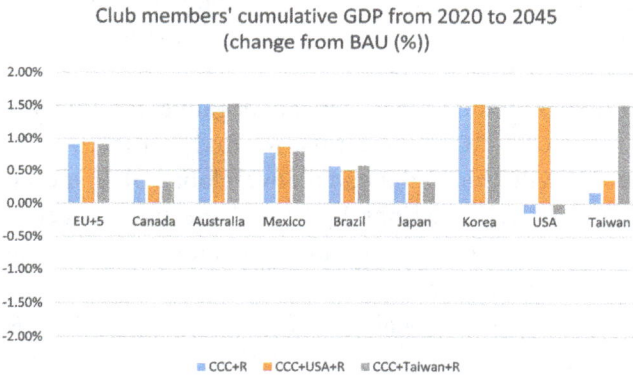

Club members' cumulative GDP from 2020 to 2045 (change from BAU (%))

Figure 3. CCC + USA or + Taiwan members' cumulative GDP from 2020 to 2045 (change from BAU (%)).

political decision for each country and these two indicators are the key factors in decision-making for each country's citizens and politicians.[4]

From Table 2, we find that

(1) Both the WCC and revenue recycling matter a lot. If every country joins the club (WCC) and implements the uniform carbon tax, the global CO_2 emissions in 2050 can be reduced by 50% from BAU, but with a negative impact on the global GDP. According to the interactive version of figure SPM.1 in the Summary for Policymakers of the IPCC Special Report on Global Warming of 1.5°C (Masson-Delmotte *et al.*, 2018), under this scenario, the likely range of human-induced warming relative to 1850–1900 in 2050 is between 1.4°C and 2.1°C.[5] On the other

---

[4]The E3ME model provides many annual indicators of environmental and macroeconomic impacts such as CO_2 and GHG emissions, the structure of electricity generation and energy consumption for the environment and energy, and the output, employment and wages by industry, the price index, income distribution, and GDP by components up to 2050.
[5]See https://apps.ipcc.ch/report/sr15/fig1/index.html.

hand, revenue recycling can make a massive difference in the impacts on GDP at the price of a slight increase in CO_2 emissions (1.7%). The global cumulative GDP compared to BAU is changed from a negative impact without tax revenue recycling to a positive one with tax revenue recycling. The reason why tax revenue recycling makes such a significant difference is that the lump-sum transfers trigger more consumption and investment.

(2) Under the CCC scenario with only seven most likely countries, global emissions only decrease by 1.6 billion tCO_2 by 2050. Similarly, we can see that there is a negative effect on GDP without revenue recycling. However, with revenue recycling, the negative effect could be offset and even have positive effects not only on GDP but also on CO_2 emission reduction. It is a win–win solution. This 3.8% reduction of global CO_2 emissions is because an economically more robust CCC makes nonmembers weaker economically and emit fewer emissions to offset the slight increase of members' emissions. The same reason applies to the following two comparisons of scenarios, i.e., with versus without trade sanctions and with versus without international transfers.

(3) The effects of trade sanctions found here are negative. By comparing the scenarios of CCC + R and CCC + R + 10% tariffs, we find that the tariff sanctions against nonclub members will reduce the cumulative global GDP and, at the same time, increase the global CO_2 emissions by 4%. These trade effects are conceivable. More details of the trade effects on different countries' GDPs can be observed in Fig. 1 and are discussed later.

(4) The introduction of international transfers under the CCC + R + 0.1 Transfer scenario, compared to the scenario of CCC + R, will increase the cumulative global GDP and reduce global CO_2 emissions slightly. It is a win–win solution too.

Thus, we find that the four parts of climate clubs have quite different environmental and macroeconomic impacts. Imposing a uniform carbon tax accompanied by domestic revenue-neutral recycling can result in a significant improvement in both environmental and macroeconomic performance being achieved. This confirms the double dividend hypothesis. We would like to point out that it is imperative to invest in the R&D of negative emissions technologies that have not yet surfaced to reduce the remaining 50% of CO_2 emissions under the WCC scenario to meet the net-zero target by 2050. However, the negative effects of trade sanctions make it unclear if they can fulfill their intention of inducing more club participation. Finally, the win–win solution of international transfers can not only enhance equality among club members but also induce more likely participation in the climate club.

Figure 1 shows the macroeconomic impacts for club members and nonmembers under different scenarios of the Core Climate Club in terms of the change in a country's cumulative GDP from BAU. We can see that a uniform carbon tax without any revenue recycling would result in a negative impact on GDP for all club members. However, some nonmembers, such as Taiwan, may benefit, and some others, such as the USA, may lose out. This may be because Taiwan, Korea, and Japan are close trade partners and competitors. A carbon tax with revenue recycling, on the other hand, would offset the negative effect and even have a positive effect on GDP for all members and most of the nonmembers.

By comparing the two scenarios of CCC + R and CCC + R + 10% tariffs, the effects of trade sanctions found here are negative for most members except for EU + 5, Canada and Mexico. Among the negatively impacted members, Japan and Korea are hit the most. Their cumulative GDPs become smaller than those under BAU. This may be because they are trade-intensive and resource-poor countries. Therefore, they encounter a lot of losses from higher import and export prices.

As for nonmembers, some of them do benefit from trade sanctions, for example, the USA, Russia, China, and Taiwan. No research in the past has found positive effects of trade sanctions on nonmembers. The reason why some nonmembers would benefit from trade sanctions is because of the trade substitution effect. Nonmembers would trade with other nonmembers instead of with their original trade partners that are now CCC members and impose trade sanctions. Since the E3ME model uses historical data to simulate the future, it can simulate the possible transformation of trade relationships.

Transfers would improve equality between club members. Among these seven countries, Mexico and Brazil would benefit from transfers since they are the recipients of international transfers. All benefactors would suffer from a slight loss of GDP. EU + 5 as a whole benefits slightly because many countries among EU + 5 are recipients, such as Sweden and France, and some others are benefactors, such as Luxembourg, Netherlands, and Germany. As for nonmembers, some of them do suffer from transfers, for example, the USA, China, and India, the three major economies of the world. As mentioned above, when we discussed Table 2, the transfer makes the CCC as a whole stronger and makes nonmembers weaker economically.

By comparing the three scenarios of CCC + R, CCC + R + 10% tariffs and CCC + R + 0.1 Transfer, we can see the negative impacts of trade sanctions and the minor effects of transfers. Since Japan and Korea are essential members of the CCC and both of them suffer a lot from trade sanctions, we drop trade sanctions from the policy instruments of the CCC and arrive at the scenario of CCC + R + 0.1 Transfer, which makes a big difference for Japan, Korea, Brazil, and Mexico.

From Fig. 2, we can see that the resources transferred among club members as GDP percentages have changed a lot between 2020, 2030, and 2040. For Brazil and Mexico, the transfers received are going to be gradually reduced significantly. One primary reason is the penalty of their expected high population increases when we use the 2020 population as the basis to calculate the number of transfers (see Footnote 2). On the other hand, the forecast that the population in Korea will reach its peak in 2029 and its 2040 population is going to be smaller than that in 2020 may be one of the reasons that make Korea pay more transfers as a percentage of GDP in 2040.

To examine whether or not nonmembers are likely to join the CCC, we compare two completely different economies, namely, the USA and Taiwan, as examples of different kinds of nonmembers. The USA is a large open, but less trade-intensive, economy. In 2019, exports and imports accounted for 11.7% and 14.6% of GDP in the USA, respectively. By contrast, Taiwan is a resource-poor, small, open and highly trade-intensive economy. In 2019, exports and imports accounted for 64.0% and 53.6% of GDP in Taiwan, respectively. Figure 3 compares the macroeconomic impacts for club members under three different compositions of the club, CCC, CCC+USA, and CCC+Taiwan, with a carbon tax and revenue recycling. We can see that both the USA and Taiwan have strong incentives to join the CCC because the GDPs of both the USA and Taiwan are much higher on joining the CCC than under the CCC scenario. Furthermore, most of the seven CCC members would benefit more if the USA or Taiwan joined the CCC, except that Australia, Canada, and Brazil would lose a little if the USA joined the CCC. In addition to the macroeconomic impacts presented in Fig. 3, any new member joining the CCC would certainly reduce its CO_2 emissions significantly and make a nontrivial contribution to global CO_2 emissions reduction (see the last two rows in Table 2).

6. Conclusion

Previous studies reveal several aspects of climate clubs as a possible solution to facilitate international cooperation on climate change. However, there are still several gaps that need to be filled. We address these gaps using the E3ME-FTT model to assess the impacts of alternative climate club structures that include four parts: a uniform carbon tax, domestic revenue-neutral recycling, trade sanctions, and transfers. The beauty of the E3ME-FTT model is that it can simulate the scenarios with these alternative structures up to 2045.

We find that, first, domestic revenue-neutral recycling is crucial, which has not been considered in the quantitative assessment literature on climate clubs. The impact on GDP compared to BAU will change from a negative one without tax revenue recycling to a positive one with tax revenue recycling. Secondly, some nonmembers may be punished by trade sanctions. In addition, some other nonmembers may benefit from it, such as the USA, Russia, China, and Taiwan. No research in the past has found positive effects of trade sanctions on nonmembers. The reason why some nonmembers will also benefit from trade sanctions is that they will trade with other nonmembers instead of their original trade partners who are now CCC members and who may impose trade sanctions.

Furthermore, some of the CCC members, such as Japan and Korea, will also be harmed by trade sanctions, since they are resource-poor and trade-intensive countries and will suffer large losses from higher import and export prices. Thus, we suggest that trade sanctions not be used as part of the design of the climate club. The positive macroeconomic and environmental impacts of a carbon tax with tax revenue recycling are sufficiently significant to keep the existing members and induce nonmembers to participate in the club. Third, international transfers will improve equality between club members, discourage population increase, and induce the participation of non-members. By contrast, the members who pay the transfers will only suffer a slight loss of their GDP. Therefore, we suggest that international transfers are a proper part of the design of the climate club.

Acknowledgments

We are grateful to Academia Sinica for funding this research through the research project AS-KPQ-106-DDPP. We also thank Hector Pollitt, Unnada Chewpreecha, and Pim Vercoulen at Cambridge Econometrics for their helpful technical support of the E3ME model, and Professor Robert Mendelsohn at Yale University and Professor Soocheol Lee at Meijo University for their useful comments and suggestions.

Appendix A

We calculate the amount of the transfers according to the following formula:

For benefactor country i, whose CO_2 emission per capita is higher than the club average

$$
\begin{aligned}
\text{TRANSFER}_i &= \text{total carbon tax revenue}_{\text{higher than avg}} \times 10\% \\
&\times \frac{(CO_2 \text{ per capita}_i - CO_2 \text{ per capita}_{\text{club avg}}) \times \text{population } (2020)_i}{\sum_i (CO_2 \text{ per capita}_i - CO_2 \text{ per capita}_{\text{club avg}}) \times \text{population } (2020)_i}.
\end{aligned}
$$

For recipient country j, whose CO_2 emission per capita is lower than the club average

$$
\begin{aligned}
\text{TRANSFER}_j &= \text{total carbon tax revenue}_{\text{higher than avg}} \times 10\% \\
&\times \frac{(CO_2 \text{ per capita}_{\text{club avg}} - CO_2 \text{ per capita}_j) \times \text{population } (2020)_j}{\sum_j (CO_2 \text{ per capita}_{\text{club average}} - CO_2 \text{ per capita}_j) \times \text{population } (2020)_j}.
\end{aligned}
$$

References

Baker III, JA, M Feldstein, T Halstead, NG Mankiw, HM Paulson Jr, GP Shultz, T Stephenson and R Walton (2017). The conservative case for carbon dividends. Climate Leadership Council.

Barrett, S (2003). *Environment and Statecraft: The Strategy of Environmental Treaty-making.* Oxford University Press.

Böhringer, C, JC Carbone and TF Rutherford (2016). The strategic value of carbon tariffs. *American Economic Journal: Economic Policy*, 8(1), 28–51.

Burke, J, R Byrnes and S Fankhauser (2019). How to price carbon to reach net-zero emissions in the UK. *Grantham Research Institute on Climate Change and the Environment and Centre for Climate Change Economics and Policy*, London School of Economics and Political Science, London.

Camuzeaux, J, T Sterner and G Wagner (2020). India in the coming 'Climate G2'? *National Institute Economic Review*, 251, R3–R12.

Chewpreecha, U and TY Lee (2015). The distributional effects of low carbon policies in Japan and South Korea. In *Low-carbon, Sustainable Future in East Asia: Improving Energy Systems, Taxation and Policy Cooperation*, pp. 171–189.

Ekardt, F and A von Hovel (2009). Distributive justice, competitiveness, and transnational climate protection: One human-one emission right. *Carbon & Climate Law Review*, 2009(1), 102–113.

Hovi, J, DF Sprinz, H Sælen and A Underdal. (2019). The club approach: A gateway to effective climate cooperation?*British Journal of Political Science*, 49(3), 1071–1096.

IPCC (2018). Global warming of 1.5°C. Retrieved August 23, 2019, from https://www.ipcc.ch/sr15/.

Klenert, D, L Mattauch, E Combet, O Edenhofer, C Hepburn, R Rafaty and N Stern (2018). Making carbon pricing work for citizens. *Nature Climate Change*, 8(8), 669.

Lee, SC, H Pollitt and SJ Park (eds.) (2015). *Low-carbon, Sustainable Future in East Asia: Improving Energy Systems, Taxation and Policy Cooperation*, pp. 29–41. England: Routledge.

Lee, S, H Pollitt and K Fujikawa (eds.) (2019). *Energy, Environmental and Economic Sustainability in East Asia: Policies and Institutional Reforms*. England: Routledge.

Martin, N and JC van den Bergh (2019). A multi-level climate club with national and sub-national members: Theory and application to US states. *Environmental Research Letters*, 14(12), 124049.

Masson-Delmotte, TWV *et al.* (2018). IPCC, 2018: Summary for Policymakers. In *Global Warming of 1.5 C*. An IPCC Special Report on the impacts of global warming of 1.5 C above pre-industrial levels and related global greenhouse gas emission pathways, in the context of strengthening the global. World Meteorological Organization, Geneva, Tech. Rep.

Mercure, JF (2012). FTT: Power: A global model of the power sector with induced technological change and natural resource depletion. *Energy Policy*, 48, 799–811.

Nordhaus, WD (1975). Can we control carbon dioxide? IIASA, Laxenburg.

Nordhaus, WD (1977). Strategies for the control of carbon dioxide (No. 443). Cowles Foundation for Research in Economics, Yale University.

Nordhaus, W (2015). Climate clubs: Overcoming free-riding in international climate policy. *American Economic Review*, 105(4), 1339–1370.

Paroussos, L, A Mandel, K Fragkiadakis, P Fragkos, J Hinkel and Z Vrontisi (2019). Climate clubs and the macro-economic benefits of international cooperation on climate policy. *Nature Climate Change*, 9(7), 542–546.

Pollitt, H, E Alexandri, U Chewpreecha and G Klaassen (2015). Macroeconomic analysis of the employment impacts of future EU climate policies. *Climate Policy*, 15(5), 604–625.

Smith, A, U Chewpreecha, JF Mercure and H Pollitt (2019). EU climate and energy policy beyond 2020: Is a single target for GHG reduction sufficient? In *The European Dimension of Germany's Energy Transition*, pp. 27–43. Springer.

Sprinz, DF, H Sælen, A Underdal and J Hovi (2018). The effectiveness of climate clubs under Donald Trump. *Climate Policy*, 18(7), 828–838.

Torpman, O (2019). The case for emissions egalitarianism. *Ethical Theory and Moral Practice*, 22(3), 749–762.

Victor, DG (2006). Toward effective international cooperation on climate change: Numbers, interests and institutions. *Global Environmental Politics*, 6(3), 90–103.

Victor, DG (2011). *Global Warming Gridlock: Creating more Effective Strategies for Protecting the Planet*. Cambridge University Press.

CHAPTER 9

WILLIAM NORDHAUS: A PIONEER

CHARLES D. KOLSTAD

Stanford University
366 Galvez St. Stanford, CA 94305, USA

University of California
Santa Barbara, CA 93106, USA

National Bureau of Economic Research
Cambridge, MA 02138, USA

Resources for the Future
Washington, DC 20036, USA
ckolstad@stanford.edu

Keywords: Climate; IAM; integrated assessment; Nordhaus; environmental economics; global warming; climate change; greenhouse gases; DICE; Nobel Prize.

Two roads diverged in a wood, and I —
I took the one less traveled by,
And that has made all the difference.

Robert Frost

1. Introduction

Bill Nordhaus comes from a long line of pioneers. His grandfather, Max Nordhaus, left Germany in the early 1880s and decided to settle in the wild west of the US. He chose Las Vegas, New Mexico, towards the end of the Santa Fe Trail, and arrived about the same time as the railroad and Billy the Kid. New Mexico was a "territory" then — it would not be a state for another 30 years — and its population was about 100,000 (about one person per square mile). It is hard for me to contemplate choosing to move from Europe to a remote unpopulated and undeveloped location on the frontier, with all the risks (but also opportunities) such a move entailed. Bill's father was another pioneer, in natural resource and Native American law but particularly in skiing. After serving in the famous United States Army's Tenth Mountain Division in the mountains

This chapter was originally published in Climate Change Economics, Vol. 11, No. 4, December 2020, published by World Scientific Publishing, Singapore. Reprinted with permission.

of Italy in World War II, Robert Nordhaus led the establishment of major ski areas in the Western US (Albuquerque and Santa Fe).[1]

Which is why it is perfectly appropriate to view Bill Nordhaus as following in his father's and grandfather's footsteps, as a pioneer, but in economics and the environment rather than business or law. The mark of a pioneer is choosing a path regardless of how popular it might be or how many naysayers may doubt your choice. Climate change was not a popular academic subject in economics when Bill Nordhaus started working on it in the early 1970s, nor for many decades afterwards. But he persevered, as pioneers do, and the result has been controversial but enormously important. In this short paper, I will relate how Bill Nordhaus has inspired many who work at the interface of climate change and economics/policy, including myself, and why, in my mind, he well deserves the accolades he has received.

2. Early Years

If we journey back to the first half of the 1970s, the big issues were the environment, energy and global cooling. In the US, the Environmental Protection Agency had just been established and several keystone laws enacted: the Clean Air Act, the Clean Water Act and the Endangered species Act. The Supersonic Transport (SST) was proposed but due to anticipated cooling effects of adding particulates to the stratosphere, the US project was canceled; Europeans proceeded with the Concorde. And then of course the Energy Crisis happened, first with the US scarcity of natural gas in 1972, followed by the OPEC oil embargo and price rise of 1973.

These were heady times, though global warming/heating was not high on the public's agenda.[2] Resource scarcity was of concern, both to the general public but also to the economics profession. In that context, a newly minted economics professor, William Nordhaus, deemed global heating a sufficiently interesting and important topic to devote a good deal of attention to the subject. He published a paper in the *American Economic Review* in 1974 suggesting with rudimentary analysis that global warming was a long-term problem associated with continuing to use fossil fuels (Nordhaus, 1974). Shortly thereafter, he published a working paper (Nordhaus, 1975), presenting an economic model of the phenomenon, followed by a shorter article in the *American Economic Review* (Nordhaus, 1977) documenting some of that analysis. As an indication of the profession's interest in the economics of global heating, my first encounter with Prof. Nordhaus was in an energy policy session at an Operations

---

[1]Robert J. Nordhaus led a law firm representing many of New Mexico's Native American tribes and also was responsible for developing the ski area and aerial tram on Sandia Peak in Albuquerque (Source: Obituary of Robert J. Nordhaus, *Albuquerque Journal*, February 26, 2007). Undoubtedly, Prof. Nordhaus' mother and other grandparents were also pioneers, though I know less about them.

[2]Although global warming was identified as a major threat in 1965 by President Lyndon Johnson's Science Advisory Committee (The White House, 1965), it was not a major public issue in the US until the late 1980s, ironically due to a particularly hot summer in the US in 1987. Technically, negative global warming (global cooling) was of public concern in the early 1970s, though that was associated with fine particulates in the stratosphere rather than greenhouse gases.

Research Society conference around 1980. It was one of those sessions where the number of people in the audience was almost exactly the same as the number of presenters. We were both presenting — he on climate change.

Energy was a big issue through the remainder of the 1970s, with oil prices skyrocketing following the Iranian revolution in 1978–79. In an omnibus energy bill responding to this crisis, the Energy Security Act of 1980, the United States Congress, among many other things, mandated and funded a study by the National Academy of Sciences (NAS) of the problem of rising carbon dioxide levels in the atmosphere. Prof. Nordhaus (and Thomas Schelling, another Nobel-winning economist) served on that committee. That committee reported in 1982 and within that report one can see the further solidification of Prof. Nordhaus' economic analysis of climate change, building on his work in the 1970s (Nordhaus and Yohe, 1983). One conclusion was that simple models of the evolution of the global climate provided estimates of the global mean temperature that were roughly the same as estimates from much more complex general circulation models. This finding paved the way for the relatively simple model of climate incorporated into Prof. Nordhaus' DICE model, developed a decade later. A second result of the NAS study was a detailed analysis of future climate based on a model developed by Prof. Nordhaus and a former student, now an economics professor, Gary Yohe.

Most of the rest of the 1980s was dominated by other public policy issues, such as the threat of losing the stratospheric ozone layer, the collapse in the price of oil and the collapse of the Soviet Union. But just as sure as the sun rises, the summer of 1987 was hotter than usual (as summers sometimes are) and public opinion turned to the greenhouse effect as a likely culprit. Prof. Nordhaus was ready for this and within a few years had developed and published the elements of a model of climate change (Nordhaus, 1991).

3. The Pioneer Strikes Gold

The climate policy debate in political circles in the late 1980s and early 1990s seemed to be between the "do something quickly" and "let us wait and see". This is the perfect setup for an economic model that helps clarify how the various important factors in the debate interact and generate wellbeing and costs — where the tension is between acting and waiting. The research Prof. Nordhaus and collaborators had been doing for the nearly two decades prior was ready to be integrated into a policy model.

In 1992, DICE was born (Nordhaus, 1992, 1993) — the Dynamic Integrated Climate Economy model. DICE is a policy model, for assessing various climate policies; but DICE is also a teaching and learning tool, to understand how the climate system interacts with the economy. DICE builds on Prof. Nordhaus' previous two decades of work in the economics of climate change and as such is a natural extension of that work. Unlike most empirical policy models, DICE is elegantly and simply structured, based on a classic textbook model of economic growth, driven by technological change, population change and capital accumulation. Most empirical models used for policy analysis are complex and more of a black box than a transparent representation of the economic system. It is not so with DICE. The structure of DICE is clean. All of the detail and nuance goes into parameterizing the model. The model is structured to be pedagogically elegant and easy to understand. That elegance and simplicity is not without cost — a legitimate criticism is that it is better as a pedagogic tool than a forecasting tool.

Structurally, DICE is a single-sector global optimal growth model with climate added. Productive activities generate greenhouse gases as a side product, which can be reduced at a cost in consumption units. Similarly, accumulating greenhouse gases increase global temperature and that temperature increase has an economic cost, also measured in consumption units. Net consumption is gross production less investment, expenditures on emissions abatement and damage from increases in temperature. The objective of the model is to maximize the net present value of the utility of net consumption over the next few centuries. Although the model is solved as an optimization problem, its solution is equivalent to how a functioning market would operate, provided greenhouse gases are efficiently managed (by the fundamental theorems of welfare economics).

One of the unique features of DICE is that parameterizing different aspects of the model has generated research projects that stand on their own. Perhaps the best example of that is the climate change damage function in DICE, which spawned a classic paper in estimating the damages from climate change: the Ricardian analysis of how climate affects agricultural land values in the US (Mendelsohn *et al.*, 1994).

Perhaps the most remarkable aspect of DICE is how amenable it is to investigating fundamental questions in the management of greenhouse gases. Both Prof. Nordhaus and others unaffiliated with him have used DICE to explore many issues — spatial disaggregation, uncertainty, learning, the nature of discounting, different assumptions about abatement costs and damage from warming, whether innovation will solve the greenhouse problem, using demand-revealing mechanisms to solve the international agreement problem and many other fundamental issues associated with managing greenhouse gas emissions. We explore two of those questions in the next section.

4. DICE as a Pedagogic Tool

One of the remarkable aspects of DICE is that it is so amenable to exploring a wide variety of aspects of the climate problem. Because at its core it is a theoretically simple model, it is amenable to being modified to explore a wide range of issues. Over the past 30 years, a wide variety of researchers, including Prof. Nordhaus, have availed themselves of this powerful feature of DICE. For example, Google Scholar shows well over 2000 cites to the original documentation of DICE in the journal *Resource and Energy Economics* and the 1994 book which also documents the DICE model, Nordhaus (1994). Of course, not all of those cites are from papers which involve an extension or application of DICE, but many are.

To illustrate the intellectual flexibility of DICE, consider two different applications — one concerning evolving uncertainty about climate change and the other concerning the promise of innovation and technological change to magically solve the climate change problem.

4.1. *Induced innovation*

Representing market failures in an equilibrium model often presents a challenge. In the case of the externality of greenhouse gas emissions, it is solved by internalizing the externality — incorporating aggregate damage into the objective function. Research and development (R&D) which leads to innovation also involves a market failure, but one that is more difficult to represent in an equilibrium model such as DICE. Yet for many, innovating out of the climate problem is the ultimate route out of the problem. The idea is that if fossil fuel prices rise or carbon becomes priced, innovators will respond, developing new technologies which are "carbon-saving". The problem with including induced innovation (innovation induced by higher prices on carbon) is that there are market failures. The main market failure is that the incentive to innovate is diluted by the inability of the innovator to capture all of the benefits of the innovation. This leads to an under-investment in carbon-saving innovation.

Despite this challenge, a number of researchers have included a representation of innovation in DICE. In particular, Nordhaus himself (Nordhaus, 2002) modified DICE to include a representation of induced innovation. The question he asked is, "Instead of assuming exogenous decline in the emissions output ratio in a society, what happens if R&D investments are endogenous choice variables which spur the decline?" He noted (and documented) that the social rate of return on R&D is about four times the private rate of return for corporate R&D, mostly due to spillovers. Finally, he also assumed innovation acts by reducing the emissions-output ratio. He thus modified DICE by replacing the exogenous downward slope over time in the emissions-output ratio to changes driven by R&D. The knowledge stock grows with additional R&D but also depreciates over time. Finally, there is a production function which converts the level of R&D and the stock of knowledge into a rate of reduction of the emissions-output ratio. To have the model choose levels of R&D similar to what the market would choose, he inflated the cost of R&D by a factor of four. In runs of his R&DICE model, he found that induced innovation performs worse than the standard DICE model in reducing emissions, a result that is troubling for those who argue that innovation will solve the climate change problem. Popp, a former student of Nordhaus, had taken a similar approach to representing induced innovation within DICE (Popp, 2004), coming up with the ENTICE model (i.e., Endogenous Technical Change Integrated Climate Economy model). Popp (2004) provided more detail on parameterizing the model. Similar to Nordhaus, he found that induced innovation is disappointing in terms of "solving" the climate change problem.

4.2. *Uncertainty*

A second important issue in climate policy is uncertainty. For at least 30 years, there has been a political tension in the context of uncertainty about climate change (in fact, uncertainty for many environmental actions): act now before it is too late versus wait and learn more about the problem and then act appropriately based on better knowledge. Again, this is an issue that DICE is equipped to address, suitably adding the state of knowledge about climate change as a state variable in the model. The three questions that become relevant are: how is that state of knowledge defined, how does it evolve over time and how does changing uncertainty influence "actions" within DICE? The policy question then becomes "How does that fact that we are learning more and more about climate change influence our public decision about when to take definitive action?"

A number of people have addressed this problem [for a review of earlier work, see Kolstad (1996)]. To my knowledge, the first effort to modify DICE to address the problem was Kolstad (1994). In that work, uncertainty was over the damage from increased temperatures — it could turn out to be high or it could turn out to be low. The state of knowledge is the probability that damage is high. That state of knowledge evolves with exogenous learning. Learning about the true state of the world (0 or 1) could proceed slowly or quickly. Results showed that the more quickly uncertainty was resolved, the lower current period emissions control would be.

In a more general version of this problem, Kelly and Kolstad (1999) reformulated the DICE model as an infinite horizon model using a Bellman equation. In that paper, the uncertainty regards the climate sensitivity, one of the most important uncertain parameters in climate science and policy. The climate sensitivity is the long-run rise in global temperatures from a doubling of greenhouse gases. If it is low, the world will warm more slowly than if it is high. The state of knowledge is represented by the mean expected climate sensitivity and its variance. The assumption about learning is that we observe the changing weather and update our prior on the state of knowledge based on the weather we observe. In other works, learning takes place by observing a weather draw from the climate (which is the weather distribution) and then using Bayes Rule to update the prior state of knowledge. One of the conclusions is that learning is a very slow process, taking nearly a century to significantly resolve. A number of other authors have subsequently added to this literature on using DICE to address uncertainty and learning (e.g., Traeger, 2014).

5. Conclusions

The purpose of this paper has been to argue that William Nordhaus is a true pioneer in the economics of climate change. Other researchers have also made very important contributions, but Bill Nordhaus started it all and persevered decade after decade until the rest of us saw the importance of the problem and his work. This paper has not tried to be comprehensive but rather has suggested some of the important ways the DICE model has helped us better understand the economics of climate change. I for one am very grateful for Bill Nordhaus' leadership.

References

Kelly, D and CD Kolstad (1999). Bayesian learning, growth and pollution. *Journal of Economic Dynamics and Control*, 23(4), 491–518.

Kolstad, CD (1994). George Bush versus Al Gore: Irreversibilities in greenhouse gas accumulation and emission control investment. *Energy Policy*, 22(9), 771–778.

Kolstad, CD (1996). Learning and stock effects in environmental regulation: The case of greenhouse gas emissions. *Journal of Environmental Economics and Management*, 31(1), 1–18.

Mendelsohn, R, WD Nordhaus and D Shaw (1994). The impact of global warming on agriculture: A Ricardian analysis. *American Economic Review*, 84(4), 753–771.

Nordhaus, WD (1974). Resources as a constraint on growth. *American Economic Review*, 64(2), 22–26.

Nordhaus, WD (1975). Can we control carbon dioxide? Working Paper WP-75-63, June, International Institute for Applied Systems Analysis, Laxenburg, Austria.

Nordhaus, WD (1977). Economic growth and climate: The carbon dioxide problem. *American Economic Review*, 67(1), 341–346.

Nordhaus, WD (1991). To slow or not to slow: The economics of the greenhouse effect. *Economic J.*, 101, 920–937.

Nordhaus, WD (1992). An optimal transition path for controlling greenhouse gases. *Science*, 258(5086), 1315–1319.

Nordhaus, WD (1993). Rolling the 'DICE': An optimal transition path for controlling greenhouse gases. *Resource and Energy Economics*, 15(1), 27–50.

Nordhaus, WD (1994). *Managing the Global Commons: The Economics of Climate Change*. Cambridge, MA: The MIT Press.

Nordhaus, WD (2002). Modeling induced innovation in climate change policy. In *Technological Change and the Environment*, A Grübler, N Nakićenović and WD Nordhaus (eds.). Abingdon, UK: Routledge.

Nordhaus, WD and G Yohe (1983). Future carbon dioxide emissions from fossil fuels. In *Changing Climate: Report of the Carbon Dioxide Assessment Committee*, pp. 87–153. Washington, DC: National Academy Press.

Popp, D (2004). Induced innovation and energy prices. *American Economic Review*, 92(1), 160–180.

Traeger, CP (2014). A 4-stated DICE: Quantitatively addressing uncertainty effects in climate change. *Environmental and Resource Economics*, 59, 1–37.

The White House (1965). Restoring the quality of our environment. Report of the Environmental Pollution Panel of President's Science Advisory Committee, Washington, DC.